Radiation and Public Perception

ADVANCES IN CHEMISTRY SERIES **243**

Radiation and Public Perception
Benefits and Risks

Jack P. Young, EDITOR
Oak Ridge National Laboratory

Rosalyn S. Yalow, EDITOR
Veterans Affairs Medical Center

Developed from a symposium sponsored
by the Divisions of Nuclear Chemistry and Technology,
Chemical Health and Safety, and
Environmental Chemistry, Inc.
at the 203rd National Meeting
of the American Chemical Society,
San Francisco, California,
April 5–10, 1992

American Chemical Society, Washington, DC 1995

Library of Congress Cataloging-in-Publication Data

Radiation and public perception: benefits and risks / Jack P. Young, editor, Rosalyn S. Yalow, editor.

    p.   cm.—(Advances in chemistry series, ISSN 0065-2393; 243)

Developed from the symposium Radiation and Society, held at the April 1992 national meeting of the ACS, San Francisco.

Includes bibliographical references and index.

ISBN 0-8412-2932-5 (cloth).—ISBN 0-8412-3037-4 (pbk.)

    1. Radiation—Health aspects.

    I. Young, Jack P., 1929- . II. Yalow, Rosalyn S. (Rosalyn Sussman), 1921- . III. Series.

RA569.R33   1994   94—35190
363.17'99—dc20   CIP

The paper used in this publication meets the minimum requirements of American National Standard for Information Sciences—Permanence of Paper for Printed Library Materials, ANSI Z39.48-1984. ∞

PRINTED IN THE UNITED STATES OF AMERICA

1994 Advisory Board

Advances in Chemistry Series
M. Joan Comstock, *Series Editor*

FOREWORD

The ADVANCES IN CHEMISTRY SERIES was founded in 1949 by the American Chemical Society as an outlet for symposia and collections of data in special areas of topical interest that could not be accommodated in the Society's journals. It provides a medium for symposia that would otherwise be fragmented because their papers would be distributed among several journals or not published at all.

Papers are reviewed critically according to ACS editorial standards and receive the careful attention and processing characteristic of ACS publications. Volumes in the ADVANCES IN CHEMISTRY SERIES maintain the integrity of the symposia on which they are based; however, verbatim reproductions of previously published papers are not accepted. Papers may include reports of research as well as reviews, because symposia may embrace both types of presentation.

ABOUT THE EDITORS

 JACK P. YOUNG received an undergraduate degree from Ball State University in 1950 and a Ph.D. in analytical chemistry from Indiana University in 1955. He has been involved in investigations relating to nuclear or radioactive materials for almost 40 years at the Oak Ridge National Laboratory (ORNL). Initially, he was connected with the molten salt nuclear reactor program in its various aspects. He developed the windowless cell technique for doing absorption spectroscopy in molten fluoride salts and carried out spectral studies of actual fuel melt. He was involved in spectroscopic studies with a group at the Transuranium Research Laboratory at ORNL. This group prepared and identified a number of oxides, halides, and other compounds of berkelium, californium, and einsteinium. As a result of these studies, they were the first to prepare and characterize many compounds of these elements. More importantly, they developed and demonstrated that in the radioactive decay of these compounds, the oxidation state of the progeny was controlled by the oxidation state of the parent, and the coordination of the daughters in the solid state was controlled by the structure of the host compound. Some interesting speculations were made in the area of chemical synthesis and structure as the result of these observations.

He was a member of a group that developed resonance ionization spectroscopy (RIS) with the attendant single-atom detection capability. Because of this technique, Young was associated with another group that literally counted cesium atoms (fission fragments) coming from the fissioning of individual californium-252 atoms. They also made shape-deformation measurements of the spontaneous-fission isomer americium-240m by related laser techniques. Currently, he is developing mass spectrometric applications of RIS and evaluating other applications of lasers to analytical determinations.

From his various research involvements he has developed a respect for radioactivity, as opposed to a fear or antagonism. He collaborated with Rosalyn S. Yalow to generate the symposium on which this book is based. Young is a member of the American Chemical Society and Sigma Xi and is a fellow of the American Association for the Advancement of Science. He is married to Jean Kennedy Young, and they have five children: James, Mark, David, Timothy, and Karen.

ROSALYN S. YALOW received an A.B. in physics and chemistry from Hunter College in 1941 and an M.S. (1942) and Ph.D. (1945) in nuclear physics from the University of Illinois. She has received 53 honorary doctorates from universities in the United States and abroad, including Princeton University, Johns Hopkins University, Washington University, and Columbia University. In 1977 she was awarded the Nobel Prize in Physiology or Medicine for the development of radioimmunoassay, a method that is used in laboratories around the world to measure hundreds of substances of biologic interest in blood and other body fluids. In 1988, Dr. Yalow received the National Medal of Science from President Reagan. She joined the Bronx Veterans Administration Medical Center in 1947 and has been a Veterans Administration Senior Medical Investigator since 1972. Currently, she is the Solomon A. Berson Distinguished Professor-at-Large of the Mount Sinai School of Medicine, City University of New York, and Emeritus Distinguished Professor-at-Large of the Albert Einstein College of Medicine, Yeshiva University of New York.

She is a member of the National Academy of Sciences and of the American Academy of Arts and Sciences. In 1986, her laboratory was designated as a Nuclear Historic Landmark by the American Nuclear Society. Dr. Yalow has received more than 50 awards and prizes, including the first Veterans Administration William S. Middleton Award for Medical Research, the Albert Lasker Basic Medical Research Award, the Gairdner Foundation International Award, the Lilly Award and Banting Medal of the American Diabetes Association, the Koch Award of the Endocrine Society, the Van Slyke Award of the American Association of Clinical Chemistry, and the G. van Hevesy Nuclear Medicine Pioneer Award. She has served as president of the Endocrine Society and as a member of numerous advisory agencies and editorial boards. She was married to the late Dr. Aaron Yalow and has two adult children, Benjamin and Elanna.

CONTENTS

ENVIRONMENTAL AND OCCUPATIONAL EXPOSURE

PREFACE

M EDIA PRESENTATIONS OFTEN MISREPRESENT subject matter in the field of radiation. The public is given one side of the story that is not completely based in the best consensus of scientific thinking. This book and the symposium on which it is based share a purpose: to promote scientific knowledge about radiation. Often, the audience receives segments of information that generate attention but not necessarily knowledge. A simple example of such partial use of facts is found in the change of the name of "nuclear magnetic resonance," an extremely useful medical imaging tool, to "magnetic resonance." The latter name was fostered by several medical organizations in the early 1980s. Although other reasons existed, a major impetus for the change was to remove the term "nuclear" from the name of the procedure. Magnetic resonance is a shorter name, but nuclear magnetic resonance is a more accurate term. The shortened version does not challenge society to understand the term nuclear but serves to perpetuate the lay person's irrational fear. The nuclear part of the name has nothing whatsoever to do with anything hazardous; it has to do with the atomic nucleus, a harmless and actually necessary bit of matter that makes up our bodies; the average human body contains on the order of 10^{27} atomic nuclei. Other, more inane examples could be given, such as the caution label on some packages of Welsbach mantles, which reads in part, ". . . this product contains a chemical known to the state of California to cause cancer."

The meeting had its origins in an oral presentation by Rosalyn S. Yalow several years ago. The interrelationship of society and radiation was developed in her presentation: the facts and fictions, the known and the unknown. In the symposium, this theme was enlarged by speakers who presented their experimental data, thoughts, and conclusions on many aspects of radiation and public perception. Most of the speakers have developed chapters for this book. The chapters are written by prime researchers in their fields, not authors surveying the works of others. Of the few speakers who did not contribute to this volume, the talk by Leonard Sagan entitled "The Science and Politics of Low Dose Risk Estimation" deserves mention. Interested readers can find his views presented in *Low Dose Irradiation and Biological Defense*, published by Elsevier Science Publishers. In addition to chapters by the speakers, this volume includes chapters on such topics as nuclear waste disposal and the use of iodine-131 in medical treatments and tests.

This book consists of four sections. The first few chapters present an overview and general discussions about radiation and public perception. The second section deals with the health aspects of radiation, good and bad alike. The third section covers the segments of the world population who were exposed to grossly excessive amounts of radioactivity. Whatever the reason for this exposure, study of these people provides important knowledge for future societies. Chapters on the life-long study of people in Japan subjected to the ravages of nuclear weapons are included. A chapter on studies of the Chernobyl nuclear power plant is also included, although it is too early to gain definitive information. The last section deals with research related to environmental and occupational exposure to radiation.

This book includes studies by several researchers pertaining to the impact of radiation on particular segments of society. Chapters deal with the public perception of nuclear risks, storage of nuclear wastes, irradiation of foods, and future considerations of nuclear power. These chapters present the definitive views of leaders in their fields. Varied scientific points of view are brought together in a consideration of the many facets of the subject.

Other books deal with this subject matter, some with a scientific basis and some without. Interested readers should investigate these other sources, mindful of the level of scientific facts considered. With sufficient reading should come knowledge; with knowledge, understanding; with understanding, proper and responsible action for society's progress.

ACKNOWLEDGMENTS

THE EDITORS ACKNOWLEDGE the financial assistance of the following organizations in supporting the symposium and therefore the publication of this book. Help was received from the American Chemical Society through the Pedagogical Symposia Subcommittee of the Committee on Science and Divisional Activities and from three divisions of the American Chemical Society: the Division of Nuclear Chemistry and Technology, the Division of Chemical Health and Safety, and the Division of Environmental Chemistry, Inc. Financial help was also received from the Office of Epidemiology and Health Surveillance of the Department of Energy, the Division of Environmental Hazards and Health Effects of the Centers for Disease Control and Prevention, the Electric Power Research Institute, Martin Marietta Energy Systems, Inc., and Dupont Medical Products. The efforts of Karen E. Young, who created the logo used in the symposium and on the cover of this book, and the help of R. L. Mlekodaj, who assisted in creating the glossary for this book, are appreciated. Finally, without the tireless efforts of Sharon L. Lantz, neither the symposium nor the book would have been possible.

Radiation and Public Perception

Rosalyn S. Yalow

Solomon A. Berson Research Laboratory, Veterans Affairs Medical Center, Bronx, NY 10468 and The Mount Sinai School of Medicine, City University of New York, New York, NY 10029

Exposure to natural radiation is increased 10-fold over the average exposure in some regions with no detectable harmful effects. Survivors of the Hiroshima–Nagasaki bombing experienced only a 6–7% increase in cancer above that generally expected in Japan. A 20-year follow-up recorded the history of 35,000 patients who had received ^{131}I-uptake diagnostic tests that delivered 50 rem to the thyroid. The study revealed no increase in thyroid cancer among those tested for reasons other than a suspected tumor. A cooperative study of 36,000 hyperthyroid patients revealed no difference in the incidence of leukemia between those treated surgically and those treated with ^{131}I, which delivers 10 rem of total-body radiation. Environmental Protection Agency statements suggest that radon in the home causes up to 20,000 lung cancer deaths each year. Yet in the absence of smoking, lung cancer is a rare disease. Other studies demonstrate that radiation exposure is a much weaker carcinogen than the general public believes.

THE PERCEPTION OF REALITY in our world is too often confused with reality. Few subjects elicit more confusion than the popular perception of the hazards of exposure to low-level radiation and low-level radioactive wastes.

Much of the public fear of radiation has been generated by the association of radiation and radioactivity with nuclear explosions and nuclear war. The American public is so phobic on this subject that the old dream of atoms for peace, including the use of nuclear reactors

0065–2393/95/0243–0001$08.00/0

for power production and even the use of radioactive materials in biomedical investigation and clinical medicine, is threatened. Recently the public was frightened by Environmental Protection Agency (EPA) statements that 5000–20,000 lung cancer deaths yearly are caused by the naturally occurring radioactive gas, radon, that seeps into our homes. Everyone should be provided with accurate information concerning the health effects, particularly the possible carcinogenic effects, associated with low doses of ionizing radiation delivered at low dose rates.

Natural Background Radiation

Environmental background radiation from natural sources has always been our principal source of radiation exposure. This exposure arises from three sources: cosmic radiation, our self-contained radionuclides (primarily the naturally radioactive isotope of potassium, ^{40}K, with a half-life of over a billion years), and the natural radioactivity of soil and building materials.

The average whole-body natural background radiation dose in the United States, not taking into account exposure to radon and its daughters, is considered to be about 0.1 rem per year. However, those living in the Rocky Mountain regions of the United States receive on the average approximately an additional 0.1 rem each year, primarily from increased cosmic ray exposure.

Contrary to popular belief, the cancer rates in the seven states with the highest background radiation are about 15% lower than the average cancer rate for the rest of the country (1). A 1982 study (2) that took into account possible complicating factors such as industrialization, urbanization, and ethnicity appeared to confirm the reduced cancer mortality in high-altitude regions. Data such as these might suggest a protective effect resulting from excess radiation delivered at low dose rates, although other factors might be considered. Nonetheless, had the cancer incidence or mortality rate been higher than average in the Rocky Mountain states, some would have unequivocally declared radiation effects to be the causative agent.

Natural background exposure can vary as much as 10-fold in various regions of the world. Cosmic radiation increases at higher elevations. In some areas of Brazil, India, China, and perhaps elsewhere people live on naturally highly radioactive soils. Epidemiologic studies (3–5) in several of these regions have revealed no evidence of deleterious health effects associated with the marked increases in natural background radiation.

Radiation from Nuclear Weapons

Much of our knowledge about the biological effects of radiation is based on studies of people who were exposed at high doses and high dose rates. The 82,000 survivors of the Hiroshima–Nagasaki bombings were the largest group ever exposed to virtually instantaneous high doses of whole-body radiation. In this group, whose exposure averaged 27 rem, the incidence of malignancies through 1978 was only about 6% greater than would have occurred without the radiation exposure (6). That is, 4500 cancer deaths would have been expected in an unexposed population; an additional 250 cancer deaths, 90 of which were leukemia, were assumed to be a consequence of the radiation. A continuing study of this group 4 years later (7) showed about the same percentage of increase in cancer over the expected levels for Japan. The increased incidence of leukemia was most visible because it peaked at 5–9 years after the bombing and decreased thereafter.

A significant increase in breast cancer was also detected. However, this increase was observed (7) only in women who were under 39 at the time of the bombing. Thus concern about induction of breast cancer by X-ray examinations should not contraindicate the recommended use of mammography in screening for breast cancer because such screening is generally recommended only for women in the older age group.

The cumulative X-ray exposure associated with the medical follow-up of atom bomb survivors has not been taken into account in considering their radiation exposures. For instance, it has been estimated (8) that the cumulative medical doses to the stomach of a participant might be as high as 50 rem.

Medical Diagnosis and Therapy

What are the known health effects associated with the use of radioactivity in medical diagnosis and therapy? Patients treated with radioiodine, ^{131}I, for hyperthyroidism are probably the largest group receiving medically administered whole-body radiation. A study (9) of 36,000 such patients from 26 medical centers (22,000 were treated with ^{131}I and most of the rest with surgery) revealed no difference between the two groups in the incidence of leukemia. The average bone-marrow dose was estimated to be about 10 rems, more than half of which was delivered within 1 week after administration. The follow-up for the ^{131}I-treated group averaged 7 years, quite long enough to have reached the peak incidence for leukemia, as was determined from the Hiroshima–Nagasaki experience. A follow-up (10) of these same patients 3 years later also revealed no difference in leukemia rates between the two groups. A subsequent study (11) of another group of over 10,000 ^{131}I-treated hyperthyroid patients who were followed for an average of 15 years also failed to show an increase in leukemia.

By the time the radioimmunoassay of thyroid-related hormones was introduced in 1968 for use in the diagnosis of thyroid disease, radioiodine uptake had been the method of choice for more than 20 years. The average thyroidal dose received during such uptake studies was on the order of 50 rem. By 1968 between one and three million people had received ^{131}I-uptake studies for the diagnosis of thyroid disease in our country alone. There has been no systematic follow-up of these people in the United States. However, thyroid cancer remains a rare disease, accounting for only about 1000 of the 500,000 cancer deaths annually.

In Sweden there was a 20-year follow-up (12) of about 35,000 patients, 5% of whom were under 20 at the time of ^{131}I diagnostic testing between 1951 and 1969. Their average thyroidal dose was 50 rem. Among those who were studied for reasons other than a suspected tumor, the ratio of the observed number of thyroid cancers to that expected for a control group was 0.62.

Does this ratio of significantly less than 1.0 suggest a protective effect from tracer doses of [131]I? The increased risk of patients who received diagnostic tracer tests because of a suspected thyroid tumor was greater than twofold; apparently in some cases their physicians' suspicions were justified. In discussing this study the National Research Council states (13) that 50 thyroid cancers were found in the ^{131}I group, compared with an expected 39.37 cases, to yield an overall standardized incidence ratio of 1.27. It states that the results of these studies do not support the conclusion that diagnostic doses of ^{131}I significantly increase the risk of thyroid cancer (13, p 289). However, the Committee on Biological Effects of Ionizing Radiation (BEIR V) report fails to mention the difference in incidence ratios between those tested simply for function and those tested because they were suspected to have a thyroid tumor. This omission leaves the false impression that there was an increase in thyroid cancer in the ^{131}I group, although the increase may not be statistically significant.

Occupational Exposure

What is the evidence for radiation-related malignancies among radiation workers? A 1981 report (14) of mortality from cancer and other causes among 1338 British radiologists who joined radiologic societies between 1897 and 1954 revealed time-related differences. Among those who entered the profession before 1921 the cancer death rate was 75%

higher than that of other physicians. However, those who entered radiology after 1921 had cancer death rates comparable to those of other professionals. Although the exposures of the radiologists were not measured, it has been estimated that those who entered the profession between 1920 and 1945 could have received accumulated whole-body doses on the order of 100 to 500 rem.

Another large group of radiation workers studied (15) was composed of men in the American armed services who were trained as radiology technicians during World War II and who subsequently served in that capacity for a median period of 24 months. A description of their training included the statement that, "During the remaining two hours of this period the students occupy themselves by taking radiographs of each other in the positions taught them that day". The students did not receive a skin erythema dose nor did they show a drop in white count, monitoring procedures that are insensitive to acute doses less than 100 rem. From what we now know, it is likely that these technicians received as much as 50 rem or more during their training and several years of service. Yet a 29-year follow-up (16) of these 6500 radiology technicians revealed no increase in malignancies when compared with a control group of similar size consisting of army medical, laboratory, or pharmacy technicians.

There is no doubt that early radiation workers were highly exposed. This situation resulted partly from ignorance of the potential hazards associated with high doses of irradiation and partly from the absence of convenient monitoring devices. Methods for monitoring radiation were developed largely because of the Health Physics program associated with the Manhattan Project that had the responsibility for developing the atom bomb. At present the only people receiving unmonitored occupational radiation exposure are airline crews. A round-trip flight between New York and Tokyo results in each passenger and crew member receiving a dose of about 0.02 rem from the increased cosmic radiation at flying altitudes. Thus crew members who make one such flight a week receive yearly radiation doses of about 1 rem. This level is greater than the exposures received by 90% of monitored radiation workers.

Military Nuclear Test Sites

Atomic Veterans. Considerable publicity has been given to problems of the so-called atomic veterans. Caldwell et al. (17) reported an increased incidence of leukemia among 3200 men who had participated in Operation Smoky, a nuclear explosion at the Nevada Test Site in 1957. Stimulated by this report, the Medical Follow-up

Agency of the National Research Council (18) studied the mortality and causes of death of a cohort of 46,186 participants, about one-fifth of the total number of participants in one or more of five atmospheric nuclear tests. The reanalysis confirmed that among the participants at Operation Smoky the standardized mortality ratio (SMR) for leukemia was 2.5; that is, there were 10 observed leukemia deaths although only 3.97 would ordinarily be expected.

Only one of those 10 leukemia patients had received an exposure in excess of 3 rem. For all other cancers the SMRs were less than 1.0. Among participants in Operation Greenhouse at a Pacific test site in 1951, with a cohort size of almost 3000, the expected leukemia mortality was 4.43; only one case was observed, which gives an SMR of 0.23. For the other malignancies the numbers involved are much larger and the SMRs are in the range of 0.7 to 0.9. For the entire cohort of 46,000, the SMR for all malignancies is 0.84 and for leukemia it is 0.99. The excess SMR for leukemia at Operation Smoky and the equivalently decreased SMR at Operation Greenhouse are typical aberrations attributable to small-number statistics.

Could an increase in leukemia have been predicted at Operation Smoky? Because only one of the veterans with leukemia was reported to have received more than 3 rem, the probability of observing a true increase in leukemia would require a gross underestimate of the radiation dose received by the participants. A committee (chaired by Merril Eisenbud for the National Research Council) reviewed the methods used to assign radiation doses to service personnel at nuclear tests and concluded that the methods were reasonably sound. However, the committee concluded (19) that doses assigned to the test participants were probably somewhat higher, not lower, than the actual doses received. This report also reviewed a number of studies that estimated radiation exposure from internally deposited radionuclides and concluded that these did not add significantly to the external exposure.

Civilians Exposed to Fallout. Other published reports describe increases in malignancies among civilians exposed to fallout from nuclear testing. In 1979 Lyon et al. (20) reported that leukemia mortality in children had increased in those Utah counties receiving high levels of fallout from the atmospheric nuclear testing conducted in 1951–1958, compared to mortality in low-fallout counties and in the rest of the United States. In the 1944–1950 and 1959–1975 periods the leukemia mortality in the so-called high-fallout regions was considerably lower than in the rest of Utah and the United States. In addition, the sum of childhood malignancies (leukemia plus other cancer deaths) appears to follow a generally downward trend from 1944 to 1975. The drop in the high-fallout counties was somewhat greater than in the low-fallout counties, although if the standard deviations had been included the differences would not have been significant.

Lyon's paper (20) was criticized in the same and later issues of the journal by several biostatisticians (21–23). In general, their objections were related to the apparent underreporting or misdiagnosis in the earlier cohort and to errors in small-sample analysis. For instance, Bader (22) presented a year-by-year listing of leukemia cases in Seattle–King County, which has a larger population than the Southern Utah counties, and noted that there were only two cases in 1959 and 20 in 1963 among the 217 cases reported from 1950 to 1972. Thus, a 10-fold difference in annual incidence rates when the number of cases is small simply represents statistical variation.

A more recent estimation (24) of external radiation exposure of the Utah population, based on residual levels of ^{137}Cs in the soil, showed that the mean individual excess exposure in what Lyon deemed to be the "high-fallout counties" was 0.28 rad, compared to 0.42 rad in the "low-fallout counties". Even in Washington County, the region in which the fallout arrived the earliest (less than 5 h after the test), the estimated exposure to its 10,000 population averaged only 1.12 rads. This exposure level is quite comparable to natural background radiation in that region over a 10-year period.

Nuclear Reactor Accident

Much has been written about the consequences of the Chernobyl reactor accident in April 1986. The Chernobyl-type RMBK 1000 reactor differs from those used outside the Soviet Union for power production in that it resembles an early military design intended for the production of weapons-grade plutonium. The Chernobyl reactor therefore had a relatively unprotected roof through which plutonium-enriched fuel could be unloaded. The radioactive plume emerged through this unprotected roof. The explosions in the reactor resulted from complete disregard of safety procedures. Furthermore, according to a Soviet report (25), the system that would have caused the reactor to shut down of its own accord in the event of a problem had been disabled. In contrast, Western nuclear power reactors are completely enclosed in a sealed containment structure that is designed to contain the products of a severe accident for an appreciable length of time.

We may never know the complete medical consequences of this accident in the Soviet Union. However a 1991 report (26) analyzed the cumulative effective dose equivalent resulting from internal exposure to the radiocesium fallout in Austria, where the effective dose equivalent was the highest in Western Europe. The specific activity of ^{40}K in human muscle tissues obtained at autopsy averages about 110 Bq/kg. In contrast, the specific activity of the ^{137}Cs peaked at about 80 Bq/kg and of ^{134}Cs peaked at about 30 Bq/kg. Both remained elevated for only about 1.5 years, whereas the ^{40}K remains at its level continuously. Thus, the integrated radiocesium dose of the 4 years after Chernobyl averaged 25 mrem, compared to 68 mrem observed for ^{40}K over the same period.

Radon in the Home

Levels of Exposure. The last issue to be discussed is the significance of radon and its daughters in the home. By the 1930s it was appreciated that miners (in particular, uranium miners) had an increased incidence of lung cancer that was presumably due to high concentrations of radon and its daughter products in the mine air. Large-scale use of uranium commenced during World War II and considerable concern was expressed about the deleterious consequences of radiation exposure. Soon thereafter steps were taken to improve ventilation in the mines.

Indoor radon caused relatively little concern until December 1984, when a worker set off radiation alarms on his way into Pennsylvania's Limerick nuclear power plant. It was recognized then that radon contamination in homes in the Reading Prong area of Pennsylvania and New Jersey, and perhaps elsewhere in the country, might exceed levels currently permitted in mines. Subsequently the EPA publicized its estimate that as many as 20,000 of the 140,000 annual lung cancer deaths in the United States are caused by home exposure to radon and its radioactive daughters. Is this estimate reasonable?

Effect of Smoking. Let us consider what the lung cancer death rate in the United States was before cigarette smoking became common. According to American Cancer Society statistics (27), the male and female age-adjusted annual lung cancer death rates were about 4 and 2 per 100,000, respectively, in 1930; in 1988 the rates had risen to 75 and 29 per 100,000, respectively. Because there is no reason to anticipate a sex-linked difference in lung cancer, the 1930 female rate was probably closer to the true lung cancer rate in nonsmokers. Was there a marked underdiagnosis of lung cancer among women in 1930? This is not likely because the rate increased only slowly until 1960, when the effects of smoking among women after World War II resulted in a continuously steeper rise in their lung cancer death rates.

Furthermore, among Mormons in Utah who have access to very good medical care, the age-adjusted lung cancer death rate for women in 1967–1975 was only 4.7 per 100,000 and that for men was 27 per 100,000 (*28*). Because of religious beliefs, Mormons are supposed to abstain from smoking, alcoholic beverages, and even caffeine-containing drinks such as coffee and cola. Although the incidence of lung cancer for Mormon males was less than one-half of that for other American males during the same period (*27*), it does suggest that not all Mormons abstain from smoking. Therefore some Mormon females may have been smokers or exposed to passive smoking. Thus it is not unreasonable to conclude that, in the absence of smoking, the lung cancer death rate should be no more than 2–3 per 100,000.

In the high-background area in China (*3*) the natural radioactivity in the soil increases radiation exposure from radon progeny to more than twice as high as that in the control area studied (*29*). However, the age-adjusted lung cancer death rate in the high-background district is 2.7 per 100,000, compared to 2.9 per 100,000 in the control area. Thus the lung cancer death rates in both groups, in spite of the difference in radon-related exposures, were comparable to rates found among American women before the smoking era.

A 1990 report (*30*) of residential radon and lung cancer among women in New Jersey suggests a trend among light smokers of increasing lung cancer as radon levels increase. However, there was no trend of increasing lung cancer among lifetime nonsmokers with increasing home radon exposure. The data also suggested (*30*) a striking decrease in lung cancer in heavy smokers (>25 cigarettes per day) with increased home radon exposure. However, there were so few cases with high radon exposures that the conclusions had no statistical significance.

The evidence for lung cancer among nonsmokers associated with radon in the home environment is very weak. Even among nonsmoking miners, lung cancer is not found (*31, 32*) among those exposed to less than 1000 times the 70-year indoor levels that the EPA estimates (*33*) would result in between 1 and 5 lung cancer deaths among 100 so exposed. Smoking is such an overwhelming cause of lung cancer that variation in smoking patterns tends to obscure any possible effect of radon exposure. A study (*34*) of 2668 lung cancer patients found that only 134 were nonsmokers. Thus the reported American Cancer Society statement that 85% of lung cancer in the United States is caused by smoking is certainly an underestimate. It is likely that at present smoking accounts for 95% of the deaths from lung cancer in the United States. Furthermore, most cancerous lesions in nonsmokers are located in the deeper portions of the lung (*34*). Lung cancer attributable to the radon daughters would be expected to be found in the bronchial epithelium.

The EPA and the media fail to make it clear that the lung cancers they list as radon-associated are linked to the assumed multiplicative carcinogenic effect of smoking and radon daughters. The emphasis should be on reducing or eliminating smoking, not on testing for radon and its daughters. On the other hand, radiation physicists and others who are involved in radon testing have developed a lucrative business and would not want to deemphasize the role of radon in lung cancer.

Conclusions

This brief review reflects the lack of reproducible studies that unequivocally demonstrate harmful effects of radiation delivered at low doses and dose rates. A 1975 report dealing with radiation protection philosophy (35) stated unequivocally that:

> The indications of a significant dose-rate influence on radiation effects would make completely inappropriate the current practice of summing of doses at all levels of dose and dose-rate in the form of total person-rem for purposes of calculating risks to the population on the basis of extrapolation of risk estimates derived from data at high doses and dose-rates . . .

We must communicate this message to governmental agencies, to the media, and to the society in which we live. Failure to do so contributes to a radiation phobia that impacts on the beneficial roles of radiation and radioactivity in medical diagnosis and therapy, as well as in other applications such as nuclear power.

References

1. Frigerio, N. A.; Stowe, R. S. In *Biological and Environmental Effects of Low-Level Radiation*; International Atomic Energy Agency: Vienna, Austria, 1976; pp 385–393.
2. Amsel, J.; Waterbor, J. W.; Oler, J.; Rosenwaike, I.; Marshall, K. *Carcinogenesis* **1982**, *3*, 461–465.
3. High Background Radiation Research Group, China. *Science (Washington, D.C.)* **1980**, *209*, 877–880.
4. Barcinski, M. A.; Abreu, M.-D. A.; DeAlmeida, J. C. C.; Naya, J. M.; Fonseca, L. G.; Castro, L. E. *Am. J. Hum. Genet.* **1975**, *27*, 802–806.
5. Gopal-Ayengar, A. R.; Sundaram, K.; Mistry, K. B.; Sunta, C. M.; Nambi, K. S. V.; Kathuria, S. P.; Basu, A. S.; David, M. *Evaluation of the Long-Term Effects of High Background Radiation on Selected Population Groups*

on the Kerala Coast; Peaceful Uses of Atomic Energy, 1972; pp II:31–51.

6. Kato, H.; Schull, J. *Radiat. Res.* **1982**, *90*, 395–432.
7. Preston, D. L.; Kato, H.; Kopecky, K. J.; Fujita, S. *Radiat. Res.* **1987**, *111*, 151–178.
8. Kato, K.; Antoku, S.; Sawada, S.; Russell, W. J. *Br. J. Radiol.* **1991**, *64*, 720–727.
9. Saenger, E. L.; Thoma, G. E.; Tompkins, E. A. *JAMA, J. Am. Med. Assoc.* **1968**, *205*, 855–862.
10. Saenger, E. L.; Tompkins, E.; Thoma, G. E. *Science (Washington, D.C.)* **1971**, *171*, 1096–1098.
11. Holm, L. E.; Hall, P.; Wiklund, K.; Lundell, G.; Berg, G.; Bjelkengren, G.; Cederquist, E.; Ericsson, U. B.; Hallquist, A.; Larsson, L. G.; Lidberg, M.; Lindberg, S.; Tennvall, J.; Wicklund, H.; Boice, J. D., Jr. *J. Natl. Cancer Inst.* **1991**, *83*, 1072–1077.
12. Holm, L. E.; Wiklund, K. E.; Lundell, G. E.; Bergman, N. A.; Bjelkengren, G.; Cederquist, E. S.; Ericsson, U. B. C.; Larsson, L. G.; Lidberg, M. E.; Lindberg, R. S.; Wicklund, H.; Boice, J. D., Jr. *J. Natl. Cancer Inst.* **1988**, *80*, 1132–1138.
13. *Health Effects of Exposure to Low-Levels of Ionizing Radiation: BEIR V;* National Research Council. National Academy Press: Washington, DC, 1990; p 289.
14. Smith, P. G.; Doll, R. *Br. J. Radiol.* **1981**, *54*, 187–194.
15. McCaw, W. W. *Radiology* **1944**, *42*, 384–388.
16. Jablon, S.; Miller, R. W. *Radiology* **1978**, *126*, 677–679.
17. Caldwell, G. G.; Kelly, D. B.; Heath, C. W., Jr. *JAMA, J. Am. Med. Assoc.* **1980**, *244*, 1575–1578.
18. Robinette, C. D.; Jablon, S.; Preston, T. L. *Studies of Participants in Nuclear Tests:* Report to the National Research Council. Mortality of Nuclear Weapons Test Participants. National Academy Press: Washington, DC, 1985.
19. *Review of the Methods Used to Assign Radiation Doses to Service Personnel at Nuclear Weapons Tests.* Board on Radiation Effects Research. Commission on Life Sciences. National Academy Press: Washington, DC, 1985.
20. Lyon, J. L.; Klauber, M. R.; Gardner, J. W.; Udall, K. S. *N. Engl. J. Med.* **1979**, *300*, 397–402.
21. Land, C. E. *N. Engl. J. Med.* **1979**, *300*, 431–432.
22. Bader, M. *N. Engl. J. Med.* **1979**, *300*, 1491.
23. Enstrom, J. E. *N. Engl. J. Med.* **1979**, *300*, 1491.
24. Beck, H. L.; Krey, P. W. *External Radiation Exposure of the Population of Utah from Nevada Weapons Tests;* DOE/EML 401 (DE82010421); National Technical Information Service. U.S. Department of Energy: New York, 1982; p 19.
25. *Nature (London)* **1986**, *322*, 673.
26. Rabitsch, H.; Feenstra, O.; Kahr, G. *J. Nucl. Med.* **1991**, *32*, 1491–1495.
27. *Ca Cancer J. Clin.* **1992**, *42*, 28–29.
28. Lyon, J. L.; Gardner, J. W.; West, D. W. *J. Natl. Cancer Inst.* **1980**, *65*, 1055–1061.
29. Hofmann, W.; Katz, R.; Chunxiang, Z. *Health Physics* **1986**, *51*, 457–468.
30. Schoenberg, J. B.; Klotz, J. B.; Wilcox, H. B.; Nicholls, G. P.; Gil-del-Real, M. T.; Stemhagen, A.; Mason, T. *J. Cancer Res.* **1990**, *50*, 6520–6524.

31. Saccomanno, G.; Huth, G. C.; Auerbach, O.; Kuschner, M. *Cancer* **1988,** *62,* 1402–1408.
32. Roscoe, R. J.; Steenland, K.; Halperin, W. E.; Beaumont, J. J.; Waxweiler, R. J. *JAMA, J. Am. Med. Assoc.* **1989,** *262,* 629–633.
33. *A Citizen's Guide to Radon: What It Is and What To Do about It*; OPA-86–004; U.S. Environmental Protection Agency; Government Printing Office: Washington, DC, 1986.
34. Kabat, G. C.; Wynder, E. L. *Cancer* **1984,** *53,* 1214–1221.
35. *Review of the Current State of Radiation Protection Philosophy*; National Council on Radiation Protection Report No. 43; National Council on Radiation Protection and Measurements:, Washington, DC, 1975; p 4.

RECEIVED for review August 7, 1992. ACCEPTED revised manuscript December 11, 1992.

2

Public Perception of Radiation Risks

William R. Hendee

Medical College of Wisconsin, Milwaukee, WI 53226

The evolution of advanced civilization has yielded works of art and science, complex financial and political systems, and technology-driven societies such as the United States. Yet as the sophistication of these societies has increased, human self-perception has diminished. One consequence of this suppressed self-image has been a growing distrust of science and certain technologies such as nuclear energy and radiation. This apprehension has been nurtured by the news and entertainment media and has partially compromised the benefits that these technologies offer. Realization of these benefits requires a restoration of self-confidence in our ability to use technologies beneficially.

THE EPIC OF HUMAN HISTORY is marked by marvelous artistic and technological achievement. From a beginning in which existence demanded a constant struggle against the elements, other life-forms, and physical and emotional deprivation, humans have created sophisticated societies, immense metropolises, architectural marvels, artistic masterworks, and wondrous music. Complex languages have evolved to record our legends and history and to express the inner needs and yearnings of the human spirit. The development of physical science and technology has led to methods of mass production, techniques to harness the atom's energy, programs of space exploration, and efforts to build increasingly powerful computers and control systems. Future technological opportunities include advances in electronic communications and networking that may ultimately permit people to function independently of where they are in space and time. Developments in the biological sciences have paralleled those in the physical sciences.

0065–2393/95/0243–0013$08.00/0

For example, emergence of the biomedical sciences in this century has led directly to improved health, more effective medical diagnosis and therapy, and reduced morbidity and mortality from disease and injury.

In the opening sequence of *2001: A Space Odyssey*, apes discover how to use bones as tools and weapons. Finally an ape hurls a bone into the sky, and instantly the bone is transformed into a spaceship piloted by the ape's descendants. This juxtaposition captures in one moment the awesome evolution from primitive beings into creative geniuses and productive contributors to society and civilization. Humans have become a highly structured, technologically driven species, and in the process have created advanced societies with opportunities for intellectual and artistic growth unburdened by the struggle for daily survival. As individuals and as groups we may not always recognize or take advantage of these opportunities. But they are within reach, at least for many of us.

Paradox of the Modern Era

The modern era of science and technology is the product of fertile minds provided space and time for rigorous thinking and scientific experimentation within societies that value creativity and reward ingenuity. But a paradox of the modern era is that it tends to devalue the characteristics of creativity and ingenuity that are essential to its development and survival. Part of this devaluing process is a communal self-image that has less stature and magnificence than the vision of the human species that was characteristic of earlier eras. The poet Archibald MacLeish describes this fallen self-image as the Diminishment of Man (*1*). MacLeish asks,

> And yet we cannot help but wonder why—why the belief in man has foundered *now*—precisely *now*—now at the moment of our greatest intellectual triumphs, our never-equaled technological mastery, our electronic miracles. Why was man a wonder to the Greeks—to Sophocles of all the Greeks—when he could do little more than work a ship to windward, ride a horse, and plough the earth, while now that he knows the whole of modern science he is a wonder to no one—certainly not to Sophocles' successors and least of all, in any case, to himself?

In 1977 William Perkins stated to the assembly of the International Science Fiction Association (*2*), "For some reason, man—the maker of cathedrals and symphonies, who has walked on the moon and split the atom—sees himself today only as excessively numerous, grasping polluters of what would otherwise be an idyllic world. We are just that semicivilized throng that surrounded Charlton Heston in *Soylent Green*," or "the insensitive, polluting, destructive characters that Bruce Dern fought—even killed—when they threatened to destroy his forest in *Silent Running*."

The diminished view of human nature is endemic among those who proclaim the Age of Aquarius. It is intrinsic to the Gaia philosophy of a living earth. These belief systems and others that flirt with astrology, pantheism, and mysticism are antiscience, antitechnology movements that deprive the individual of the right to explore the unknown and harness nature's forces for the benefit of humanity. Despite their good intentions, these movements isolate the individual from self-determination and in the process prevent people from reaching their greatest potential of intellectual and artistic creativity. In this manner they diminish the individual and dehumanize the species.

Fear of Technologies

The innate desire to know and understand the world is a fundamental human characteristic that distinctly separates us from other forms of animal life. The application of knowledge to the development of new technologies is a natural consequence of this characteristic. Yet many people react to new technologies with feelings of fear and dread, especially when the technologies evolve from a relatively new and unfamiliar body of scientific knowledge. These feelings cause those who hold them to question the wisdom of exploring new frontiers of scientific knowledge and the desirability of attempting to unlock nature's secrets. This response is a direct manifestation of the diminished view of humanity that characterizes the modern era of science and technology. An excellent example of this diminished perspective is the continuing opposition to efforts to use nuclear energy and ionizing radiation for the benefit of society.

Fear of nuclear energy and radiation undoubtedly is in part a reaction to the devastation that accompanied the introduction of nuclear weapons at Hiroshima and Nagasaki. Because of World War II, development of the awesome force of nuclear weapons was kept hidden as a military secret. Revelation of this force over a course of 3 days in 1945 was a shock to almost everyone. After the temporary euphoria in America that followed the end of the war, many people began to wonder if scientists had not finally gone too far. The concern was whether scientists had not succeeded in penetrating the dark side of nature. Even some nuclear scientists had reservations, exemplified by

Robert Oppenheimer's perhaps apocryphal quote from the *Bhagavad Gita* while watching the Trinity Explosion at Alamogordo: "I am become death, the shatterer of worlds." As a scientist stated in the closing scene of the movie *Them*, "When man entered the Atomic Age, he opened the door into a new world. What we will eventually find in that new world nobody can predict."

The new world of nuclear energy was a relatively unexplored arena for science and technology. Many believed that this arena was imperfectly understood, probably unpredictable, possibly uncontrollable, and even perhaps malevolent. Some felt that it should never have been opened to human inquiry, and that with the end of the war it should have been closed and sealed forever against further exploration. Those who persist in such exploration might well be driven to unpredictable and uncontrollable behavior, as demonstrated by Peter Sellers in the 1964 movie *Doctor Strangelove*.

The Two Fictions of Nuclear Energy

With the advent of the nuclear age, a library of science fiction literature quickly developed around themes such as world dominance through control of the ultimate nuclear weapon, creation of new beings through exposure to radiation, devastation of the planet by nuclear war with few or no human survivors, and return of humanity to a primordial state of existence as a consequence of the technology of the nuclear era. Dozens of B-grade movies depicted the ability of radiation released from nuclear explosions, sources, and waste to transform various life-forms into horrendous monsters. Typical films were *The Thing* (1951), *The Beast from 20,000 Fathoms* (1953), *Them* (1954), *Godzilla* (1954), *It Came from beneath the Sea* (1955), *Tarantula* (1955), *The Black Scorpion* (1957), *Attack of the Crab Monsters* (1957), *The Beginning of the End* (1957), *Attack of the Giant Leeches* (1959), and many others. These movies were different from the science fiction films of an earlier era. The monsters were more horrendous, and they had an origin—exposure to radiation and nuclear energy.

Early in the nuclear era, efforts to countermand the negative perception of nuclear energy resulted in an intense and equally fictional account of the possibilities of this new energy source to provide immense benefits. Artificial suns to control weather, pollution-free "white cities of the future" electricity too cheap to meter, deserts made to bloom, immense reservoirs of petroleum and natural gas released by nuclear explosions, atomic "magic bullets" to cure cancer, and automobile engines as small as a human fist were predicted not only by science fiction writers but also by technocrats and government officials, especially those associated with the U.S. Atomic Energy Commission. As the unrealistic nature of these predictions became apparent over the next couple of decades, the real benefits of nuclear energy also became suspect. At the same time, the long-term health consequences of exposure to radiation were beginning to show up in the survivors of Hiroshima and Nagasaki.

To many people in the 1950s and 1960s, the risk–benefit balance of nuclear energy became increasingly tilted on the side of risk. The cold war, arms buildup, civil defense, fallout shelters, anticommunist mania, and other societal programs exacerbated fears of nuclear energy. These fears were heavily colored by the images of nuclear energy projected by the news and entertainment media. As Spencer Weart noted (3),

> Radioactive monsters, utopian atom-powered cities, exploding planets, weird ray devices, and many other images have crept into the way everyone thinks about nuclear energy, whether that energy is used in weapons or in civilian reactors. The images, by connecting up with major social and psychological forces, have exerted a strange and powerful pressure within history.

Benefits of Nuclear Energy

While the negative images of nuclear energy have been reinforced by the news and entertainment media over the postwar decades, the technology has found several applications of immense benefit to humanity. These applications include the use of reactor-produced radioisotopes for localizing potential deposits of petroleum, detecting flaws in construction materials, identifying and modifying hereditary characteristics of plants and insects, and measuring pathways of environmental pollution in the ecological sciences. Radioactive sources have been instrumental in the evolution of entire new areas of scientific knowledge, such as biochemical mechanisms of metabolism in plants and animals, molecular origins of processes such as angiogenesis and atherosclerosis, and mechanisms of transmission of neurological signals across synapses between nerve cells.

The two most widespread applications of nuclear energy yield the most immediate direct benefit. These applications are the use of nuclear energy in the generation of electricity and the use of radioisotopes produced in nuclear reactors for diagnosis and treatment of many human conditions, including cancer, cardiovascular disease, metabolic disorders, and mental illnesses. However, even these applications are not without controversy. For example, the use of nuclear energy to generate electricity has encountered so much opposition that no application for a nuclear power plant has been filed since 1977.

Perceptions of Nuclear Energy and Radiation

In 1989 Vincent Covello, David McCallum, and Marie Pavlova (4) identified the factors that lead to enhanced or lessened public concern associated with particular technologies. These factors are shown in Table I. Nuclear energy is one technology that satisfies all of the criteria for increased concern. The consequences of the Hiroshima and Nagasaki explosions; accidents at Three Mile Island and Chernobyl; delayed health effects attributed to radiation exposure; perceived inadequate understanding of the science underlying nuclear energy; awareness of the susceptibility of children and future generations to radiation-induced health effects; perceived potential uncontrollability of nuclear energy; the involuntary nature of exposure to risks of nuclear explosions and environmental radiation; a sense of fear and dread that accompanies nuclear energy; media hype with regard to radiation and nuclear energy; the dominance of untrustworthy utilities, public institutions, and governmental agencies in controlling nuclear energy; and uncertain benefits of nuclear compared with conventional energy sources for electricity generation all combine to enhance the perception of risk and the public apprehension of nuclear energy.

Recombinant DNA technology (genetic engineering) is the only other technology that comes close to nuclear energy in satisfying virtually all of the criteria for increased public concern depicted in Table I. In America this technology has not yet encountered the severity of public reaction that has afflicted the nuclear energy industry, perhaps because it is a newer technology that has not yet achieved a high level of public awareness. The situation is quite different in Germany, where terrorist threats and attacks have been directed toward molecular biologists involved in gene-transfer experiments. In America gene manipulation has received favorable media attention compared with nuclear energy. However, the fortunes of this technology could sour quickly if a few real or perceived misguided adventures were to occur. Such adventures have been predicted for several years by doomsday prophets such as Jeremy Rifkin.

The public's apprehension about nuclear energy is reflected in a 1990 survey by Paul Slovic (5) of the attitudes of technical experts and the public about selected technologies. Results are shown in Table II for technologies associated with radiation. Nuclear power is perceived by the public as extremely risky and unacceptable, whereas technical experts consider nuclear power moderately risky but acceptable in light of its benefits to society. Nuclear weapons are viewed by the public

Table I. Factors Involved in Public Risk Perception

Factor	Conditions Associated with Increased Public Concern	Conditions Associated with Decreased Public Concern
Catastrophic potential	Fatalities and injuries grouped in time and space	Fatalities and injuries scattered and random
Familiarity	Unfamiliar	Familiar
Understanding	Mechanisms or process not understood	Mechanisms or process understood
Uncertainty	Risks scientifically unknown or uncertain	Risks known to science
Controllability (personal)	Uncontrollable	Controllable
Voluntariness of exposure	Involuntary	Voluntary
Effects on children	Children specifically at risk	Children not specifically at risk
Effects on future generations	Risk to future generations	No risk to future generations
Victim identity	Identifiable victims	Statistical victims
Dread	Effects dreaded	Effects not dreaded
Trust in institutions	Lack of trust in responsible institutions	Trust in responsible institutions
Media attention	Much media attention	Little media attention
Accident history	Major and sometimes minor accidents	No major or minor accidents
Equity	Inequitable distribution of risks and benefits	Equitable distribution of risks and benefits
Benefits	Unclear benefits	Clear benefits
Reversibility	Effects irreversible	Effects reversible
Personal stake	Individual personally at risk	Individual not personally at risk
Scientific evidence	Risk estimates based on human evidence	Risk estimates based on animal evidence
Origin	Caused by human actions or failures	Caused by acts of nature or God

SOURCE: Adapted with permission from reference 4. Copyright 1989 Plenum Press.

Table II. Relative Risks of Selected Technologies as Perceived
by Experts and the Public

| Technology | Perceived Risk | |
	Technical Experts	Public
Nuclear power and nuclear waste	Moderate risk Acceptable	Extreme risk Unacceptable
X-rays	Low to moderate risk Acceptable	Very low risk Acceptable
Radon	Moderate risk Needs action	Very low risk Apathy
Nuclear weapons	Moderate to extreme risk Tolerance	Extreme risk Tolerance
Food irradiation	Low risk Acceptable	High risk? Acceptability questioned
Electric and magnetic fields	Low risk Acceptable	Not yet aware

SOURCE: Reproduced with permission from reference 5. Copyright 1990 National Council on Radiation Protection and Measurements.

as extremely risky, but tolerable because they reduce the likelihood of a nuclear attack. Radon is seen by the public as very low risk, reflecting the public apathy about this environmental hazard. Technical experts consider radon to be a moderately risky environmental hazard that requires action. Judged by the factors that exacerbate public concern in Table I, radon satisfies only a few of the criteria for increased concern. The press and some scientists have challenged the hazard of environmental radon, thereby justifying inaction on the part of many homeowners. X-rays also are viewed as very low risk, probably because X-rays have been used routinely in medicine for many years. Perhaps even more important, X-ray examinations are prescribed by physicians, to whom most people entrust the responsibility for their health and well-being. That is, most people believe that if X-rays were truly hazardous, physicians would not expose patients to them.

Communicating Risk Information

Informing people about technologies such as nuclear energy and ionizing radiation is a major challenge. Most people react to these technologies with preconceived notions that give rise to fear and worry in some applications, and to apathy and disregard in others. Dismissing these notions as irrational or inaccurate is almost always futile. Nothing about people's perceptions is recognized as inaccurate. Declaring perceptions irrational is tantamount to accusing those holding such views of being irrational. Any effort to address the benefits and risks of technologies where perceptions are involved requires that the perceptions be acknowledged from the beginning as real and deeply felt. Ignoring perceptions, or dismissing them as unrealistic or absurd, is equivalent to ignoring or dismissing the persons who hold the perceptions. Under such circumstances, barriers to communication are often erected that impede effective discourse among individuals, no matter how expert some of the individuals may be (6).

Summary

People's perceptions of risks are the product of complex interactions among many influences, including knowledge, opinions stated by esteemed and trusted persons, and impressions acquired over time from the news and entertainment media. Cultural heritage is a particularly important factor, especially when health risks such as those associated with radiation and nuclear energy are portrayed to the public in a provocative manner (7). Over the past few decades, the entertainment and news media have focused on nuclear energy as a particularly risky technology. The consequences of this portrayal remain to be evaluated in their entirety. In all likelihood this portrayal will complicate the acceptance of nuclear energy in the future to an even greater degree than in the past. As one example of this portrayal, the author recently conducted an informal survey of one publisher, Marvel comics, and identified more than 70 comic-book characters who had developed severe physical and emotional handicaps as a consequence of exposure to radiation. The message inherent in the experience of these characters is communicated effectively to young persons and to many others who are not so young.

Like other technologies, nuclear energy has the potential of contributing substantially greater benefits to society. These benefits include improved health care and the efficient production of electricity at a time when access to other resources for electricity generation are becoming increasingly problematic. This technology, like all others, has a destructive potential, a "dark side" in the vernacular of technology cynics. This potential can be harnessed and controlled for the benefit of humanity, as has been proven in medicine and the nuclear power industry over the past several decades. The greater benefits of nuclear energy will be realized, however, only if the diminished view of human nature is replaced by a perception that cherishes the fundamental ingenuity and integrity of the human species.

Expansion of existing technologies and the development of advanced new technologies require acknowledgment of the intrinsic worth of the search for knowledge and the desirability of developing new technologies as a product of expanded knowledge. They also require confidence in the inherent ability of humans to apply these technologies for the betterment of society. Many years ago we had that vision. Hopefully we will return to it in the near future.

References

1. MacLeish, A. *Riders on the Earth*; Houghton Mifflin: Boston, MA, 1978; p 23.
2. Perkins, W. *Presentation at the International Science Fiction Convention*; Miami, FL, September 4, 1977.
3. Weart, S. *Nuclear Fear: A History of Images*; Harvard University Press: Cambridge, MA, 1988; preface.
4. *Effective Risk Communication: The Role and Responsibility of Government and Nongovernment Organizations;* Covello, V.; McCallum, D.; Pavlova, M., Eds.; Plenum Press: New York, 1989; pp 3–16.
5. Slovic, P. In *Radiation Protection Today—The NCRP at Sixty Years*; Proceedings No. 11; National Council on Radiation Protection and Measurements; Bethesda, MD, 1990; p 79.
6. Hendee, W. *Postgraduate Radiology* **1991,** *11,* 164–178.
7. Hendee, W. *Health Physics* **1990,** *59,* 763–764.

RECEIVED for review August 7, 1992. ACCEPTED revised manuscript March 25, 1993.

Basic Units and Concepts in Radiation Exposures

R. L. Mlekodaj

Office of Radiation Protection, Oak Ridge National Laboratory, Building 4500S, MS-6106, P.O. Box 2008, Oak Ridge, TN 37831

Some of the most common units, concepts, and models in use today dealing with radiation exposures and their associated risks will be presented. Discussions toward a better understanding of some of the basic difficulties in quantifying risks associated with low levels of radiation will be presented. The main thrust of this chapter will be on laying a foundation for better understanding and appreciation of the chapters to follow.

T HE PROCESSES AND RISKS INVOLVED in the exposure of humans to radiation are very complicated and frequently little understood. Even though it has long been quite clear that many different effects are attributable to radiation exposure, the exact dose–response relationship for many of these effects has remained elusive. This fact is especially true for low doses, where the vast majority of human exposure actually occurs. To provide a foundation for better understanding of the ideas to be presented, this chapter will focus on some of the very basic underlying knowledge in this field. This information should be of value, especially for people who do not work with these units and concepts on a daily basis.

Units Relative to Exposure to Ionizing Radiation

Quantity of Radioactive Material. A quantity of radioactive material can be described in terms of the number of nuclear trans-

0065–2393/95/0243–0023$08.00/0

formations that result in the loss of the identity of the decaying species (disintegrations) per unit time. The most commonly used units are disintegrations per second (dps) or disintegrations per minute (dpm). One of the most commonly used units of radioactivity is the curie (Ci). One curie is defined as the quantity of any nuclide for which the disintegration rate is 3.7×10^{10} dps. A curie is a large quantity of radioactive material and will generally only be found in specialized facilities designed to handle high levels of radioactive material. Generally, one is more likely to encounter radioactive materials involving millicurie, microcurie, nanocurie, or picocurie quantities. The corresponding unit in the International System of Units (SI) is the becquerel (Bq), and is defined as 1 dps. Therefore, 1 Ci is equal to 3.7×10^{10} Bq.

Exposure. Exposure is the oldest radiation dosimetric unit still in common use. Dosimetric units are essential for quantifying exposures for biological effects experiments, controlling exposures to individuals, and so on. Exposure is only defined for electromagnetic radiation (gamma and X-rays) and is a measure of the ionization produced in dry air at standard temperature and pressure. It is the sum of all of the ions of one sign when all electrons liberated by photons in a volume element of air are completely stopped in the air, divided by the mass of air in the volume element. The unit of exposure is the roentgen (R) and is defined as

$$1 \text{ R} = 2.58 \times 10^{-4} \text{ C/kg}$$

where C is coulombs and kilograms (kg) refer to the mass of air. One roentgen of exposure corresponds to about 0.95 rad of absorbed dose (*see* the following section) in soft tissue and, due to the close numerical correspondence, leads to frequent misuse of terms (e.g., milliroentgen per hour is frequently stated when millirad per hour or millirem per hour is the correct term). There is no corresponding SI unit.

Absorbed Dose (Dose), D. The fact that the roentgen applies only to air and electromagnetic radiation created a need for a more generally applicable unit; in particular, one that could be applied to tissue. The traditional unit of absorbed dose is the rad and was developed to apply to any directly or indirectly ionizing radiation in any absorbing medium. One rad is defined as the absorption of 100 ergs/g from the radiation field. The SI unit of absorbed dose is the gray (Gy) and is defined as an absorbed dose of 1 J/kg.

$$1 \text{ Gy} \equiv 1 \text{ J/kg} = 100 \text{ rad}$$

In common usage, the absorbed dose is generally referred to as simply "dose".

Dose Equivalent (Equivalent Dose), *H*.   Dose equivalent is the basic unit of importance for radiation protection programs. Dose equivalent is defined as

$$H = DQ$$

where *H* is the dose equivalent in rems, *D* is the absorbed dose in rads, and *Q* is the quality factor. In principle, other modifying factors can also be added to the right side of this equation, but, in practice, this is rarely done. The quality factor *Q* arises from the fact that certain types of radiation produce a higher probability for stochastic effects in biological systems for equal amounts of energy absorption per unit mass (absorbed dose). In ICRP 26 (*1*), a useful empirical relationship was established between the linear energy transfer (LET), or collisional stopping power, and *Q* for charged particles in water. For simplicity in the administration of radiation protection programs, however, the National Council on Radiation Protection and Measurements (NCRP) and the ICRP have through the years made recommendations as to the assignment of *Q* for various types of radiation. The 1987 recommendations of the NCRP (*2*) are given in Table I. The most recent recommendations of the ICRP 60 use the new term "radiation weighting factor" instead of "quality factor" and advocate the use of the term "equivalent dose" instead of "dose equivalent". The SI unit for dose equivalent is the sievert (Sv) and is equal to the absorbed dose in grays times the quality factor. Thus, 100 rems is equal to 1 Sv.

Effective Dose Equivalent (Effective Dose), H_E.   In order to account for nonuniform irradiation of different organs or tissues, a quantity is defined in ICRP 26 such that a combination of different doses to different tissues can be combined in a way that is likely to correlate well with the total risk for stochastic effects. This quantity

Table I. NCRP 91 Recommended Values of *Q*

Type of Radiation	Approximate Value of Q
X-rays, γ-rays, β-particles, and electrons	1
Thermal neutrons	5
Neutrons (other than thermal), protons, alpha particles, and multiple-charged particles of unknown energy	20

is effective dose equivalent (effective dose as suggested in ICRP 60) and is defined as

$$H_E = \sum_T w_T \cdot H_T$$

where H_T is the dose equivalent in organ or tissue T and w_T is a weighting factor representing the proportion of the stochastic risk when the whole body is irradiated uniformly. The values of w_T (ICRP 26) are given in Table II. The unit of effective dose equivalent is the rem or the sievert.

The remainder of 0.30 not accounted for in Table II is assigned, at a level of 0.06, to each of five remaining organs or tissues that receive the highest dose equivalent. It is assumed that exposures of all remaining tissues can be neglected. Recommendations for new values of w_T are given in ICRP 60.

Committed Dose Equivalent (Committed Equivalent Dose), $H_{T,50}$. This quantity is the dose equivalent (equivalent dose) to an organ or a tissue T that will be accumulated over the 50 years following a single intake of radioactive material and can be expressed as

$$H_{T,50} = \int_{t_0}^{t_0+50y} \dot{H}_T(t)\, dt$$

where $\dot{H}_T(t)$ is the appropriate dose-equivalent (equivalent dose) rate and t_0 is the time of intake. The unit of committed dose equivalent is the rem or the sievert. The 50 years is intended to represent a typical working life. This quantity is probably inappropriate for one who is either very young or very old at the time of intake. ICRP 60 recommends the use of the term "committed equivalent dose" for this quantity.

Table II. Tissue Weighting Factors of ICRP 26

Tissue	w_T
Gonads	0.25
Breast	0.15
Marrow	0.12
Lung	0.12
Thyroid	0.03
Bone surfaces	0.03
Remainder	0.30

Committed Effective Dose Equivalent $H_{E,50}$. The committed effective dose equivalent (CEDE) is obtained by extension and combination of the concepts of committed dose equivalent and effective dose equivalent. The unit of committed effective dose equivalent is the rem or the sievert.

$$H_{E,50} = \sum_{T} w_T \cdot H_{50,T}$$

ICRP 60 recommends the use of the term committed effective dose for this quantity.

Annual Effective Dose Equivalent (Annual Effective Dose). The annual effective dose equivalent (AEDE) is the total effective dose equivalent from both the internal and external irradiation of tissues and organs received in 1 calendar year.

Collective Dose Equivalent, S. The collective dose equivalent is defined for a population by the following expression:

$$S = \sum_{i} H_i P_i$$

where H_i is the dose equivalent to the whole body or any specified organ or tissue to each member of a subgroup (i) with P_i members of the exposed population. The unit of collective dose equivalent is the person · rem or the person · sievert. A collective dose equivalent of 100 person · rem could be 100 persons exposed at a level of 1 rem each or 1000 persons exposed at a level of 0.1 rem each or a myriad of other combinations. In the concept of a linear, no-threshold dose–response relationship, a 100 person · rem of collective dose equivalent should correspond to the same total risk and be independent of how the dose equivalent is distributed among those exposed. This concept can be extended to other dosimetric quantities, such as collective exposure and collective effective dose equivalent.

Cumulative Dose. Cumulative is the sum of the dose received by an individual over a specified period of time. The cumulative concept can be applied to all other dosimetric quantities except those involving the "committed" concept.

Working Level. One working level is any combination of short-lived radon daughters in 1 L of air that will result in the ultimate release of 1.3×10^5 MeV of alpha particle energy. This number was chosen because it is approximately the alpha energy released from the decay of daughters in equilibrium with 100 pCi of ^{222}Rn. One working-level month is the cumulative exposure equivalent to exposure to one working level for 1 working month (170 h).

The Interaction of Ionizing Radiation with Matter

Ionizing radiation is radiation sufficiently energetic to dislodge electrons from an atom or molecule. Ionizing radiation includes X and gamma radiation, beta radiation (β^+ and β^-), alpha radiation, heavier charged atomic nuclei, and neutrons. Of these entities, the charged particles are considered directly ionizing radiation and the uncharged particles (neutrons and photons) are considered indirectly ionizing, producing directly ionizing charged particles after interacting with matter. The ionization and excitation of atoms and molecules are the primary energy loss mechanisms for charged particles. The moving charged particle imparts energy to the electrons of the medium via electromagnetic forces, resulting in ionization or excitation.

Beta particles can also lose energy by bremsstrahlung, but this mode of energy loss remains a small contribution to total energy loss below about 50 MeV. Because the mass of the beta particles is the same as the entity with which it primarily interacts (electrons of the medium), it can lose a large fraction of its energy in a single interaction and can incur large deflections in direction.

Heavy charged particles (charged particles other than electrons or positrons) are massive compared to the electrons with which they interact and therefore travel a rather straight path. Other energy loss mechanisms (nuclear interactions) generally can be ignored for heavy charged particles.

In passing through living matter, all radiation produces an ion pair for approximately every 25 eV of energy deposited, regardless of the exact nature of the tissue in question. A typical person in the United States might receive approximately 200 mrad of penetrating radiation in a year from natural background, diagnostic medical procedures, and so on. This absorbed dose of 200 mrad would translate into about 2.5 $\times 10^{12}$ ion pairs produced in each gram of tissue irradiated at that level.

The chemical changes produced by ionizing radiation in liquid water are especially relevant, because the human body is >90% water. The initial products produced by passing charged particles in pure water are H_2O^+, H_2O^* (an excited water molecule), and electrons. The characteristic time scale for this initial stage is about 10^{-15} s. These three species then react in the following way:

$$H_2O^+ + H_2O \rightarrow H_3O^+ + OH$$

$$H_2O^* \rightarrow \begin{cases} H_2O^+ + e^- \\ H + OH \\ H_2 + O \end{cases}$$

$$e^- \rightarrow e_{aq}^-$$

where the H_2O^+ interacts with a neighboring water molecule to produce a hydronium ion and a hydroxyl radical. The excited water molecule either ejects an electron to become an ion or dissociates predominately in one of two ways. The free electron is solvated by available water molecules. These reactions are generally complete by about 10^{-14} s.

Four of these products are free radicals (H, O, OH, and e_{aq}^-) and, with H_3O^+, form five reactive species as a result of the original interaction of the ionizing radiation with pure water. The O radical quickly reacts with H_2O to form hydrogen peroxide, which reacts no further.

The four remaining reactive species begin a diffusion stage, during which they may come within reaction distances of other reactive species and be consumed according to the following reactions:

$$OH + OH \rightarrow H_2O_2$$

$$OH + e_{aq}^- \rightarrow OH^-$$

$$OH + H \rightarrow H_2O$$

$$H_3O^+ + e_{aq}^- \rightarrow H + H_2O$$

$$e_{aq}^- + e_{aq}^- + 2H_2O \rightarrow H_2 + 2OH^-$$

$$e_{aq}^- + H + H_2O \rightarrow H_2 + OH^-$$

$$H + H \rightarrow H_2$$

After about 10^{-6} s, the reactive species that have survived the preceding reactions are likely to have diffused to such separation distances that further reactions are unlikely. A more complete discussion on these chemical effects can be found in reference 3.

In a biological system, these remaining free radicals will probably react with cellular material within about 10^{-3} s. Thus, in less than a millisecond, a series of physical and chemical reactions have taken place that could be expressed as a cancer, for example, 5 or 10 years later, albeit with an extremely low probability.

Effects of Ionizing Radiation

Biological effects are assumed to arise from two types of interactions, direct and indirect. The attack of free radicals, produced by ionizing radiation on DNA and resulting in mutation, is an example of an indirect effect. Radiation can also interact directly with the DNA strand, thereby causing mutations. This reaction would be an example of a direct effect. The indirect effects are believed to dominate. The biological effects of ionizing radiation can vary widely depending on dose, dose rate, type of radiation, and many other factors.

Stochastic effects of ionizing radiation are those that occur by chance or, in other words, in a statistical manner. These are primarily cancer and genetic effects in human exposure to radiation. These stochastic effects are also characterized by three traits. First of all, there is no threshold. The *likelihood* of the effect occurring is dependent on dose. Second, the severity does not depend on dose. Finally, there is no clear causal relationship. Cancer and genetic effects are caused by many agents, and the exact cause of a stochastic effect cannot be unequivocally linked to any one agent.

Most biological effects are of the nonstochastic type. Nonstochastic effects have three characteristics in common. First, nonstochastic effects exhibit threshold doses. That is, a certain minimum dose must be exceeded before that particular effect is observed. In addition, the severity of the effect is dependent on the dose. Finally, a clear causal relationship between the exposure and the resulting effect exists. For example, a person who is exposed to sunlight must be exposed above a certain level before he or she shows signs of sunburn. More exposure to sunlight will increase the severity of the sunburn, and there is also no question that the sunburn is the result of exposure to sunlight. The International Commission on Radiological Protection (ICRP) in Publication 60 (4) recommends that nonstochastic effects be referred to as deterministic effects.

Somatic effects are those effects that are manifested in the exposed individual. These could include cataracts, cancer, or acute radiation syndrome, for example. An additional type of effect is not expressed in the exposed individual but rather in subsequent generations. These effects are referred to as genetic and appear as hereditary disorders in subsequent generations. Radiation can affect any cell in the body, but only when germ cells are altered can the defective genetic information be passed on to future generations. These genetic changes can

vary from inconsequential and unnoticed to very serious handicaps in
future generations. Radiation is not the only agent that can induce
genetic aberrations, and one can compare the natural mutation rate
to that from radiation. According to the report of the Committee on
the Biological Effects of Ionizing Radiation in BEIR V (5), about 100
rad (1 Gy) of low dose rate, low linear energy transfer (LET) radiation
to the parental population will double the naturally occurring mutation
rate.

Limits for Exposure to Ionizing Radiation

In recent decades the recommendations of the ICRP and NCRP have
been used as the basis for our national standards in radiation dose
limits. Table III shows the downward trend in the recommended maximum occupational whole-body exposure as a function of time, as recommended by the ICRP and NCRP. The most recent recommendations of the NCRP and ICRP are compared in Table IV.

The major changes from previous guidance involve added recommendations for limitations to the long-term average of the annual
limit on occupational exposure. The maximum recommended occupational annual effective dose remains, however, at 5 rem for 1 year
in both the NCRP and ICRP recommendations. The NCRP has recommended that a lifetime limit of 1 rem be multiplied by age in years
in order to limit the risk that may accumulate over a working lifetime.
The ICRP has accomplished this effect by recommending an average
of no more than 2 rem/year over defined 5-year periods.

Uncertainties in Establishing Risk for Stochastic Effects

Because planned high-dose experiments cannot be carried out on human subjects and extrapolation of risks determined in animal studies

**Table III. Maximum Permissible Occupational Whole-Body Exposure
to Ionizing Radiation**

Recommended Maximum Rate	Comments
0.2 R/day (1 R/week) (50 R/year)	Recommended by ICRP in 1934 and continued in worldwide use until 1950
0.1 R/day (0.5 R/week) (25 R/year)	Recommended by NCRP on March 17, 1934, and continued in use in United States until 1949
0.3 rem/week (15 rem/year)	Recommended by NCRP March 7, 1949, and ICRP in July 1950 and continued in use until 1956
5 rem/year (0.1 rem/week)	Recommended by ICRP in April 1956 and NCRP on January 8, 1957

Table IV. Recent Recommended Dose Limits

	NCRP 91		ICRP 60	
	Annual Limit (rem) (Annual Effective Dose Equivalent)	Limitations/Clarifications	Annual Limit (rem) [Annual Effective Dose Equivalent (External) + Committed Effective Dose Equivalent (Internal)]	Limitations/Clarifications
To Limit Stochastic Effects for				
Occupational workers	5	Lifetime limit 1 rem × age	5	<2 rem/year averaged over defined periods of 5 years
Public	0.1	Up to 0.5/year can be allowed if infrequent	0.1	>0.1 rem/year allowed in special circumstances if 5-year average <0.1 rem/year
Minors/students	0.1	None	0.1	Inferred to be same as public
Embryo/fetus	0.5	Efforts to make uniform over gestation period	0.2	None
Occupational planned special exposure	10	Lifetime limit 10 rem	—	Not discussed
To Prevent Nonstochastic Effects in				
Lens of the eye	15	None	15	None
Skin	50	None	50	Averaged over any 1 cm^2
Hands and feet	50	None	50	None
All tissues and organs except lens of the eye, skin, and hands and feet	50	None	Limit on total effective dose equivalent of 5 rem/year to limit stochastic effects is considered adequate to protect other tissues and organs from nonstochastic effects	None

are very much in question, the main sources of risk data in humans are the Japanese atomic-bomb survivors, miners exposed to high levels of radon, and a few medically related therapeutic and diagnostic programs. Among these few available study groups, the Japanese atomic-bomb survivors are by far the dominant source of data used by various groups in establishing risks related to exposure to radiation. Some of the most common difficulties encountered in establishing this risk are discussed in the following sections.

Dosimetry.   The difficulty in establishing the actual dose equivalent to which the Japanese survivors were exposed is easily appreciated. The actual dose equivalents can only be established retrospectively by calculations of the atomic-bomb yields and calculation of transport modes, which can then be compared to neutron activation studies and thermal luminescence studies of roof tiles, for example. The dose equivalent estimates have varied over the years as improved studies have provided new results.

Dose–Response Relationships.   The vast majority of our definitive knowledge of dose response in humans comes from high dose rate and high dose situations, the dominant among these being the Japanese atomic-bomb survivors studies. Results from high doses delivered at low dose rates were obtained from studies of the chronic exposure of uranium miners to radon. Studies at low doses are generally inconclusive, complicated by many confounding factors and statistical variations that occur when working with small numbers. The dose–response relationships observed for the high-dose data cannot be attributed to a unique curve through the data points, as illustrated in Figure 1. The straight-line (curve A) fit in Figure 1, which passes through the origin, is referred to as the linear, no-threshold dose–response relationship and is only one of many equally good fits to the data points shown. One other possibility (curve B), a so-called linear-quadratic fit, is equally as good a fit as the linear, no-threshold fit. In fact, a situation shown in curve C, in which the response at low doses actually falls below the abscissa, cannot be excluded. This situation would be interpreted as one in which, over some range of low doses, radiation actually had a beneficial effect—the so-called "hormesis effect". Reference 6 gives in-depth information on hormesis.

All radiation protection standards for stochastic effects are based on a linear, no-threshold relationship between dose equivalent and risk, where the risk–dose equivalent relationship is derived from exposure data in high dose and high dose rate situations.

Sensitive Subpopulations.   Studies of the Japanese survivor data indicated differences among subpopulations in sensitivity to radiation.

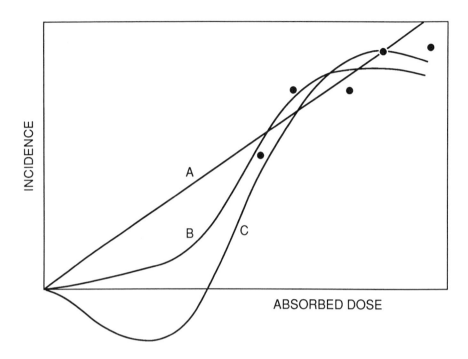

Figure 1. Dose–response relationships: (A) linear, no-threshold, (B) linear-quadratic, and (C) inclusion of a possible hormesis effect.

Increased risk to Japanese females of up to 50% was reported in BEIR V for all cancers when these women were exposed to an acute dose of low-LET radiation. For some specific cancer types, however, males showed an increased risk. The age at exposure was shown to be important. Again in BEIR V the risk for all cancers in those women exposed between 0 and 19 years of age versus those exposed between 20 and 64 years shows a threefold increase. The risk for various types of cancer was shown to be strongly dependent on nationality. The Japanese show a stomach cancer risk 10 times higher than that of Americans, but colon and breast cancer are significantly less common among the Japanese. These differences may be related to diet (*see* the subsequent discussion of cofactors). Certainly many other subpopulations, with significantly different risks for induction of cancers by radiation, exist.

Cofactors.   Perhaps the clearest example of a cofactor for risk to radiation exposure is smoking in its relation to lung cancer. In studies of several groups of uranium miners, the BEIR IV Committee's analyses indicated a significantly increased risk associated with smokers. These results are in agreement with earlier studies and can be examined in detail in the BEIR IV report (7). The significantly higher risk for stomach cancer among the Japanese due to exposure to ionizing radiation could be attributable to a cofactor related to something in the Japanese diet, but that cofactor has not yet been identified.

Incomplete Projections. The two primary types of models used to describe probability of cancer induction after exposure to ionizing radiation, the additive and multiplicative risk models, are shown in a stylized fashion in Figure 2. The simple additive model, after a characteristic latent period, gives an increased cancer death probability rate that is proportional to dose equivalent but is independent of age. The simple multiplicative model gives, again after the latency period, an increased cancer probability that is proportional to dose equivalent but is age dependent and is a multiple of the background cancer rate (which increases with age). The Japanese atomic bomb survivors are now past the point in Figure 2 where the curves of the additive and multiplicative models cross. In the past it was difficult to choose the best of these two models, but it now appears that most studies for the majority of cancers show that the multiplicative models give better fits to the data.

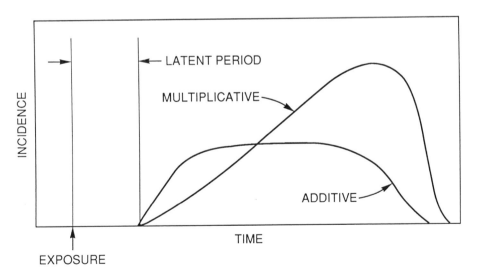

Figure 2. Comparison of the simple additive and multiplicative risk models for cancer induction.

Dose Rate Effectiveness Factor. The risk factors for cancer inductions in humans, as mentioned previously, are based primarily on high doses delivered at high dose rates. Data on dose rate factors obtained through experiments in radiobiology are summarized in NCRP Report No. 64 (8). The results of the studies suggest that, for low-LET radiation, a reduction of risk of at least a factor of 2 can be taken when the dose is delivered at a low dose rate. Studies of high-LET radiation, in some instances, show an increased efficacy for cancer induction when the LET is delivered at a low dose rate.

Unique Endpoints. In general, we have discussed cancer induction as a simple monolithic process. Realistically, each cancer type is unique and should be analyzed individually; thus, the statistical power in the limited data available is lessened. Each cancer type would be expected to have its own unique latent period as well as a unique dependence on any of the other variables we have discussed.

Statistics. The problems and pitfalls associated with the statistics of small numbers frequently act as limiting factors when attempting to establish or correlate risks associated with exposure to ionizing radiation. In a hypothetical case in which the long-term average annual rate for a particular type of cancer among a particular group was one case per year, an investigator might report a 100% increase in cancer for a year where two such cases were reported. Statistically absurd, these types of reports nevertheless find their way into the popular press, where they cause undue concern among the general population.

Healthy Worker Effect. In a large number of epidemiological studies of workers in the nuclear industry that have relatively low but well-documented radiation exposures, the results have frequently shown that these workers, on average, live *longer* than the general population. This finding is attributed to the "healthy worker effect". This effect is generally attributed to factors like (1) a higher than average educational level, (2) preselection bias in the hiring process, (3) better than average health insurance and medical surveillance, and (4) high job security.

Summary

In spite of all the problems associated with the establishment of actual risks involved in exposure to ionizing radiation that I have outlined, I do not want to give the impression that we should somehow impose more restrictive limits on radiation exposure until all these risks are more clearly understood. On the contrary, to date, epidemiological studies of workers who incur radiation exposure, within present limits, as part of their occupation have failed to detect a statistically significant increase in cancer or other health problems. Thus, I feel a large measure of conservatism has already been incorporated into our present exposure limits for ionizing radiation, and I would not recommend additional reductions in the exposure limits.

Acknowledgment

This work was managed by Martin Marietta Energy Systems, Inc. for the U.S. Department of Energy under Contract No. DE–ACO5–84OR21400.

References

1. ICRP Publication 26. *Recommendations of the International Commission on Radiological Protection*; Pergamon Press: Elmsford, NY, 1977.
2. NCRP Report 91. *Recommendations on Limits for Exposure to Ionizing Radiation*; NCRP Publications: Bethesda, MD, 1987.
3. Turner, J. E. et al. *Radiat. Res.* **1983,** *96,* 437–449.
4. ICRP Publication 60. *1990 Recommendations for the International Commission on Radiological Protection*; Pergamon Press: Elmsford, NY, 1990.
5. BIER V Report. *Health Effects of Exposure to Low Levels of Ionizing Radiation*; National Academy Press: Washington, DC, 1990.
6. Luckey, T. D. *Radiation Hormesis*; CRC Press: Boca Raton, FL, 1991.
7. BEIR IV Report. *Health Effects of Radon and Other Internally Deposited Alpha-Emitters*; National Academy Press: Washington, DC, 1988.
8. NCRP Report 64. *Influence of Dose and Its Distribution in Time on Dose–Response Relationships for Low-LET Radiation*; NCRP Publications: Bethesda, MD, 1980.

RECEIVED for review July 6, 1993. ACCEPTED revised manuscript April 1, 1994.

HEALTH EFFECTS OF RADIATION

Department of Energy Radiation Health Studies

Past, Present, and Future

Terry L. Thomas[1] and Robert Goldsmith

Office of Health, U.S. Department of Energy, Washington, DC 20585

The U.S. Department of Energy (DOE) and its predecessor agencies have been involved in research on the human health effects of exposure to ionizing radiation for more than 40 years. This long-term program began with studies of Japanese atomic bomb survivors and has since comprised studies of workers at nuclear weapons facilities, communities near those facilities, and populations potentially affected by the Chernobyl accident. In recent years the program has become more applied, with emphasis on descriptive studies and occupational health surveillance of workers, whereas the long-term, analytic research studies are being managed by the Department of Health and Human Services (HHS) under a Memorandum of Understanding (MOU). Data from these research activities are made available to the scientific community through the DOE's Comprehensive Epidemiologic Data Resource (CEDR), a public use database.

THE EARLIEST FORERUNNER of the Department of Energy (DOE) was the Manhattan Project, begun during World War II to develop atomic weapons. The 1954 Atomic Energy Act created the Atomic Energy Commission (AEC) to control the possession, use, and production of

[1]Current address: Department PMB, Uniformed Services, University of the Health Sciences, Bethesda, MD 20814

atomic energy in the United States; to conduct, assist, and foster research and development in atomic energy; and to encourage widespread participation in the development and utilization of atomic energy for peaceful purposes. In 1974 the Energy Reorganization Act placed licensing and related regulatory functions of the AEC under the auspices of the Nuclear Regulatory Commission (NRC). The Energy Research and Development Administration (ERDA) assumed the remaining AEC functions and became responsible for directing federal activities relating to research and development of various sources of energy, increasing efficiency and reliability in the use of energy, and carrying out military and nuclear weapons production activities. The DOE was created by the Department of Energy Organization Act of 1977. The DOE's mission, according to the Act, is to establish a coordinated energy policy, promote energy conservation, create awareness of energy needs in the private sector, and provide for the administration of functions of the ERDA related to nuclear weapons and national security.

DOE Radiation Health Studies: Past

Atomic Bomb Casualty Commission and Radiation Effects Research Foundation. The first epidemiologic research project supported by the DOE's predecessor agencies was initiated with the formation of the Atomic Bomb Casualty Commission (ABCC) in 1947. The ABCC was funded by the United States to study the delayed effects of external radiation exposure among survivors of the atomic bombs dropped on Hiroshima and Nagasaki during World War II. In 1975 the Radiation Effects Research Foundation (RERF) replaced the ABCC as a jointly funded binational organization. The work conducted by the ABCC and the RERF has contributed much of what is known today about the health effects of external exposure to ionizing radiation and has provided much of the basis for the subsequent development of radiation protection standards.

The epidemiologic research program of the RERF has several components. The Life Span Study is a cohort mortality study of 120,000 subjects, including bomb survivors with varying levels of exposure to external ionizing radiation and nonexposed controls (1). The vital status of the cohort is periodically updated. The Adult Health Study is a detailed medical follow-up study of a sample of 20,000 subjects from the Life Span Study who receive physical examinations and a series of clinical tests every 2 years. Studies of children born to bomb survivors are being conducted to investigate delayed effects of parental exposure, and they include an evaluation of pregnancy outcome conducted between 1948 and 1953, as well as mortality follow-up of off-spring and cytogenetic and biochemical genetics studies.

City-wide tumor registries were initiated in Hiroshima in 1957 and Nagasaki in 1958. Tissue registries were established in both cities in the early 1970s to collect and store tumor tissue samples from bomb survivors. Somatic chromosome studies have been conducted on tissues from a sample of 1200 bomb survivors. Immunology and cell biology studies examine effects of radiation on the immune system and at the cellular level (2). The in utero study examined developmental characteristics of survivors who were exposed to radiation in utero (3). A variety of special cancer studies are being carried out in addition to the main program, and a reassessment of atomic bomb radiation dosimetry is being made.

Internal Emitters Program.   The first occupational epidemiologic activity supported by DOE's predecessor agencies was a cohort study of about 4600 women who were employed in the radium dial painting industry during the early 1900s (4). The women in this study used radium-containing paint to luminize watch and clock dials and ingested radium when they used their mouths to sharpen the tips of their paint brushes. The Internal Emitters Program was begun by Argonne National Laboratory (ANL) during the 1950s to evaluate mortality and morbidity among this cohort and other groups with internal exposure to radium. The entire Internal Emitters database includes information on nearly 8000 subjects who received their exposures occupationally or through medical treatment. Measured doses are available for about one-third of the cohort. The ANL has followed the population for mortality and morbidity and maintains a large database that includes demographic information, exposure histories, and clinical examination data, as well as radiographic films for many of the subjects.

DOE Worker Health and Mortality Study.   The DOE Health and Mortality Study is composed of a variety of projects to investigate the potential long-term health effects of chronic occupational exposure to low levels of ionizing radiation (5). The study was initiated in 1964 under a contract with the University of Pittsburgh to study mortality among DOE contractor workers at the Hanford site. In 1979 the study was expanded to include additional DOE sites, and administration of the research effort was transferred to several DOE contractor organizations. Battelle Pacific Northwest Laboratory and the Hanford Environmental Health Foundation assumed responsibility for follow-up of DOE contractor workers at the Hanford site. Oak Ridge Associated Universities studied workers at the three Oak Ridge facilities, the Fernald Feed Materials Production Center, the Savannah River site, all

DOE workers exposed to 5 rem or greater in a calendar year, and workers at Manhattan Engineer District sites. The Los Alamos National Laboratory (LANL) conducted studies of workers at LANL, the Rocky Flats plant, the Mound plant, the Zia site, and workers exposed to plutonium at any DOE site.

A study of nuclear shipyard workers was conducted by the Johns Hopkins University under a DOE contract. The study included about 70,000 workers employed at eight shipyards involved in the overhaul of nuclear-powered vessels since 1957. About half of the workers were exposed externally to low levels of ionizing radiation. Because the follow-up period was short for many of the workers, the DOE plans to support continued follow-up of this population.

DOE Radiation Health Studies: Present

When Admiral James D. Watkins became Secretary of Energy in 1989, he announced a ten-point initiative that shifted the DOE's focus from production to health and safety and environmental restoration. As a part of this emphasis, Watkins chartered the Secretarial Panel for the Evaluation of Epidemiologic Research Activities (SPEERA). The panel was composed primarily of public health professionals from state and academic institutions. After conducting numerous site visits and public meetings to review the DOE's epidemiologic activities, SPEERA recommended that the DOE epidemiology program place a greater emphasis on worker and public health issues than on scientific research questions and that the DOE allow independent evaluation of its epidemiologic data. To implement these recommendations, Watkins issued six directives on March 27, 1990, ordering the establishment of a consolidated office responsible for all occupational health and epidemiologic activities at the DOE, outlining the epidemiologic functions to be carried out by this office, ordering that analytic epidemiologic research be managed externally, and calling for improved communication of epidemiologic information within the DOE complex and between the Department and the public (Figure 1). The Office of Health was formed under the Assistant Secretary for Environment, Safety, and Health as a result of these directives.

To allow independent evaluation of its epidemiologic data, the Office of Health sponsors several initiatives. Management of DOE's long-term analytic epidemiologic research studies was transferred to the Department of Health and Human Services (HHS) by Memorandum of Understanding (MOU) in December 1990. The Centers for Disease Control and Prevention (CDC) National Institute for Occupational Safety and Health (NIOSH) assumed the lead role for managing the DOE Health and Mortality Study. In addition, NIOSH is sponsoring a study

Independent Evaluation of DOE Data:
- Memorandum of Understanding
      Department of Health and Human Services
- Comprehensive Epidemiologic Data Resource
- State Health Agreement Program

Emphasis on Worker Health Issues:
- Health Surveillance Program
- Beryllium Workers Enhanced Medical Screening Program

Health Communication Program

Epidemiologic Research:
- Quick Response Descriptive Studies
- Analysis of Surveillance and Screening Data

International Programs:
- Radiation Effects Research Foundation
- Chernobyl - Health and Environmental Effects

Figure 1.   DOE radiation health studies: present.

of childhood leukemia and paternal preconception occupational exposure to ionizing radiation, similar to studies conducted in the United Kingdom (6). The CDC Center for Environmental Health is managing several off-site environmental dose reconstruction projects to evaluate the external ionizing radiation exposure of residents in communities hosting DOE facilities. The DOE and HHS chartered advisory committees to assist in setting the agenda for future analytical research projects to be carried out under this MOU.

The Comprehensive Epidemiologic Data Resource (CEDR) is a public use database developed by the DOE to share data from DOE-sponsored epidemiologic studies with the scientific community. The CEDR contains raw data and analysis files from the DOE's worker health and mortality study, including demographic, health physics, and mortality data. Individual subject information is included in the system without personal identifying information. The CEDR also contains full documentation for each of its components. In the future, data from numerous other populations will be made available through the CEDR. Proposed additions include selected data from the RERF Life Span Study, the nuclear shipyard workers study, the internal emitters program, and commercial nuclear power plant workers. A subset of the DOE's health surveillance data will also be made available through the CEDR.

A third effort to encourage independent evaluation of potential health effects associated with working in or living near a DOE facility is the State Health Agreement program sponsored by the Office of Health. Grants have been awarded to the health department or other state institution in several states that host DOE facilities. Each grant provides funds for the conduct of independent epidemiologic research and community health surveillance activities related to potential health effects of DOE operations on neighboring communities. Local advisory panels selected by the grantee oversee the conduct of activities under these grants. The panels are composed of members from the community, workers at the facility under study, scientists from a variety of disciplines, and a DOE Office of Health representative as liaison. Most panels also have a representative from the CDC. Panel meetings are open to the public, evening public meetings are often held in conjunction with panel meetings, and all materials developed under the grant are open to public scrutiny. The activities being conducted under these grants vary widely in scope and include epidemiologic feasibility studies, off-site environmental dose reconstruction projects, the development of tumor and birth defects registries, and preliminary descriptive epidemiologic investigations. The DOE presently has State Health Agreements with Florida, Tennessee, New Mexico, California, South Carolina, and Colorado.

In order to place a greater emphasis on worker health issues, the Office of Health has expanded the scope of its pilot health surveillance system which was started in 1983 at the Hanford site and the Idaho National Engineering Laboratory. The purpose of the Health Surveillance Program is to conduct continuous assessments of worker health status to identify potential workplace health hazards. The system is population based; thus, rates for specific medical conditions can be compared for various subsets of the worker population or across facilities. Morbidity data will be collected routinely and will form the basis for conducting descriptive epidemiologic studies, evaluating time lost for illness or injury, and developing registries of selected diseases and conditions. Goals are to standardize demographic, exposure, and health outcome data collection throughout the DOE complex and to make the system comprehensive to include all DOE and DOE contractor workers at all DOE sites. Demographic, medical industrial hygiene, and health physics data will be linked and integrated into the system. Worker representatives will participate in ensuring the quality and completeness of the data. Data from the Health Surveillance Program will be available to DOE occupational medical directors and epidemiologists for routine analyses. A deidentified subset of data will be made available through the CEDR.

The Office of Health is conducting an enhanced medical screening program for chronic beryllium disease among current and former workers who were exposed to beryllium. About 10,000 workers are eligible to participate in the screening examination, which consists of a physical examination, chest X-ray, and lymphocyte transformation test. The goals of the program are to identify prevalent cases of chronic beryllium disease (CBD), identify workers at high risk of CBD, remove CBD cases and high risk workers from exposure, and conduct routine medical monitoring of high risk workers. Educational materials for workers about beryllium and CBD are being developed with the assistance of the Workplace Health Fund. The program was initiated at Rocky Flats, has been expanded to include workers at the Oak Ridge Y-12 plant, and eventually will include all DOE facilities with metal fabrication operations involving beryllium.

To improve communication of epidemiologic information within the DOE complex and between the DOE and the public, a separate division within the Office of Health manages communication functions for the Office. One of the major functions of this division is to disseminate information on the health effects of energy production and use. Several types of publications are distributed throughout the DOE complex to share results from epidemiologic studies and to disseminate information about other health-related news of interest to DOE workers, including new legislation, regulations, and rules or orders. The division prepares fact sheets and other materials in lay language for the public. Tutorials on epidemiologic methods are being prepared for workers, the public, and nonepidemiologist scientists. The Office of Health is the DOE's focal point for responding to public inquiries regarding the health effects of energy generation, distribution, and use and participates in worker and public meetings.

The DOE's internal epidemiologic research program is focused primarily on health surveillance and quick response descriptive studies. Evaluations of suspected disease clusters are conducted to investigate the concerns of workers and community residents and to follow up findings from routine health surveillance activities.

The Office of Health supports several international efforts. The office provides U.S. funding through the National Academy of Sciences for the RERF. Under a bilateral agreement between the United States and the former Soviet Union, a collaborative research program on the health and environmental effects related to the Chernobyl accident was developed under DOE leadership with support from the Nuclear Regulatory Commission. Research activities are being conducted under the auspices of Working Group 7 of the Joint Coordinating Committee for Civilian Nuclear Reactor Safety. Two subworking groups were organized under Working Group 7, one concerned primarily with environmental transport of radionuclides and the other concerned with health effects of the accident. The DOE is providing technical support for an effort being coordinated by the International Agency for Research on Cancer to pool data from various countries on workers exposed to low levels of ionizing radiation. Data from the DOE Health and Mortality Study will be included in the pooled analysis.

Program support for several additional activities is provided by the Office of Health. The Radiation Emergency Assistance Center/Training Site (REAC/TS) maintains a current tally of worldwide radiation accident experiences since 1944; this tally provides a basis for evaluating the clinical course of and treatment modalities for radiation-induced injuries and the monitoring of the survivors' subsequent health status (5). A repository of information on radiation accidents of clinical and public health significance is maintained as an informational and educational resource. Support is provided for the Uranium/Transuranium Registries for investigations of the fate of radionuclides in the body (7). The Populations at Risk from Environmental Pollutants (PAREP) database contains U.S. mortality and census data broken down by census block and also has air and water quality data collected by the Environmental Protection Agency. This database will be made available through CEDR for use in ecologic-type studies.

DOE Radiation Health Studies: Future

The scope and priorities for future DOE epidemiologic studies of radiation-related health effects will be determined through an open process involving both scientific and lay input. Several initiatives appear likely, but the emphasis of the DOE epidemiology and health surveillance program probably will evolve to reflect the reconfiguration of the weapons complex, the development of alternative energy sources, and the changing roles of the national laboratories. Certainly, the Health and Mortality Study will continue under HHS management to update exposure and mortality information on previously studied populations and to include workers at DOE facilities not yet studied. Among these facilities are the Idaho National Engineering Laboratory, Lawrence Livermore National Laboratory, and Lawrence Berkeley Laboratory. Many of DOE's health surveillance activities will focus on workers involved in the cleanup of radiological wastes, and additional off-site radiation dose reconstructions will be initiated. Nevertheless, renewed epidemiologic emphasis will be placed in the area of chemical toxicity: reconstructing on-site chemical inventories, investigating nonradiological health effects, and standardizing industrial hygiene practices and the usage of job titles. In addition, the DOE will continue to take the lead federal role in research on health effects due to exposure to nonionizing radiation. In all of these areas, health research and surveillance will begin to incorporate state-of-the-art techniques for exposure assessment and early detection of morbid outcome as we move toward the era of molecular epidemiology.

References

1. Schull, W. J. Chapter 12 in this book.
2. Neel, J. Chapter 10 in this book.
3. Yoshimoto, Y.; Soda, M.; Schull, W. J.; Mabuchi, K. Chapter 11 in this book.
4. Stehney, A. F. Chapter 14 in this book.
5. Fry, S. A.; Cragle, D. L.; Crawford-Brown, D. J.; Dupree, E. A.; Frome, E. L.; Gilbert, E. S.; Petersen, G. R.; Shy, C. M.; Tankersley, W. G.; Voelz, G. L.; Wallace, P. W.; Watkins, J. P.; Watson, J. E., Jr.; Wiggs, L. D. Chapter 17 in this book.
6. Gardner, M. J.; Snee, M. P.; Hall, A. J.; Powell, C. A.; Downes, S.; Terrell, J. P. *Br. Med. J.* **1990,** *300,* 423–429.
7. Kathren, R. L. Chapter 5 in this book.

RECEIVED for review February 4, 1993. ACCEPTED revised manuscript May 24, 1993.

The U.S. Transuranium and Uranium Registries

Ronald L. Kathren

Washington State University, Richland, WA 99352

The U.S. Transuranium and Uranium Registries are unique postmortem human tissue research programs studying the biology of the actinide elements in humans. This chapter describes the history, objectives, operation, and recent scientific accomplishments of the registries and provides a listing of collaborative research activities. Findings from more than 200 autopsies are described, and new biokinetic models and parameters for plutonium and americium are given and compared with existing models of the International Commission on Radiological Protection (ICRP).

THE U.S. TRANSURANIUM AND URANIUM REGISTRIES (USTUR) are unique human tissue research programs whose origins date back more than 4 decades. In 1949, what the initiators described as "a modest program of postmortem tissue sampling at autopsy" was begun at what was then the Hanford site of the U.S. Atomic Energy Commission (AEC) (1–3). This program required the collection of samples of bone, lung, liver, and occasionally other tissues at autopsy from both Hanford workers and other residents of Richland, Washington, where most of the Hanford workers resided. Samples thus collected were radiochemically analyzed for plutonium, the goal being to evaluate sites of preferential deposition of plutonium within the body and to compare what was observed in the tissues postmortem with what was predicted on the basis of the application of biokinetic models to excretion data.

0065–2393/95/0243–0051$08.00/0

Not surprisingly, this study revealed very low levels of plutonium in the tissues of the local residents and Hanford site workers. Somewhat surprisingly, it also revealed that, at least since 1962, most of the plutonium found in the tissues resulted from fallout from nuclear weapons tests rather than occupational exposures. Although the highest individual tissue concentrations of plutonium were observed in the pulmonary lymph nodes of a worker with a history of occupational exposure, liver depositions, in general, occurred more frequently than those in the lung. Data for the bone samples collected were equivocal, and this initial report of nearly 20 years ago concluded with a plea for further investigation and collaboration with other plutonium handling facilities (1).

The initial formal presentation of the Hanford autopsy study was presented at the Seventh Annual Hanford Symposium on Biology held in Richland in May 1967, nearly 20 years after the study had begun (3). Coincidentally, the concluding paper at that same meeting was given by H. D. Bruner of the AEC Division of Biology and Medicine who, while graciously noting that the idea was not his or any one person's but rather "occurred to many men about the same time", proposed formation of a national Plutonium Registry and described progress toward that goal within the AEC (4). The primary purpose of the registry, as outlined by Bruner, would be to ensure that the details of an accidental intake of plutonium could be correlated with the subsequent health record of the worker. In addition to sketching the basic information and operating requirements for such a registry, Bruner also listed seven additional purposes, presciently noting among these that the Plutonium Registry should not be limited to plutonium but should also consider other transuranium elements (4).

History of the Registries

The USTUR thus grew out of a desire to better understand the potential health effects of plutonium incorporated into the human body, gaining not only improved understanding of the health effects of plutonium but also of the efficacy of control measures based on actual human experience. The progenitor of what is now the USTUR was formally established in August 1968 as the National Plutonium Registry by the Hanford Environmental Health Foundation (HEHF) under contract to the U.S. AEC. W. Daggett Norwood, a physician whose undergraduate education was in electrical engineering and who had figured prominently in the establishment of the medical program at the Hanford site, was appointed the founding director. He was ably assisted by Carlos E. Newton, Jr., Battelle–Northwest, a board-certified physicist who carried the title of consultant and who directed the health physics aspects of the program. Rounding out the staff was Dorothy Potter, who served as secretary and general administrative assistant.

Even before the contract award had been finalized, Philip A. Fuqua, then medical director of HEHF, sent out invitations in an effort to set up a blue-ribbon Advisory Committee to help guide the fledgling registry. The six initial Committee members included the following: J. H. Sterner, a physician from the University of Texas, and Robley D. Evans, the Massachusetts Institute of Technology physics professor noted for his studies of the radium dial painters, who were elected chairman and vice-chairman, respectively; toxicologist Lloyd M. Joshel, Dow Chemical Company; physician Clarence C. Lushbaugh, Oak Ridge Associated Universities; Thomas F. Mancuso, another physician, University of Pittsburgh; and noted medical and health physicist Herbert N. Parker, Battelle–Northwest. Wright Langham, the Los Alamos National Laboratory biophysicist whom many acknowledged as "Mr. Plutonium", was added the following year.

By the end of its first year, the registry had, with the aid of the Advisory Committee, established its basic operating procedures and begun recruitment of registrants, signing up three individuals that year. The following year, 1970, the registry's name was changed to reflect the broader programmatic concern with the other transuranic elements, as had been suggested in the prescient talk by Bruner a few years earlier (*4*).

With the passing of the AEC in 1972, support for the program was continued, first by the U.S. Energy Research and Development Administration and, most recently, by the Office of Health and Environmental Research of the U.S. Department of Energy (DOE). The U.S. Uranium Registry (USUR), was established as an administratively separate but similar program in 1978. The USUR is concerned with the biokinetics, dosimetry, and health aspects of exposure to uranium and its daughters, with emphasis on the uranium fuel cycle.

Although there were considerable overlaps in function and staff, each registry was operated as a separate program administered by a half-time physician–director, with scientific support for both registries provided by a half-time health physicist consultant from Battelle–Northwest. With the exception of tissues obtained from cases originating at the DOE Rocky Flats Facility, which were analyzed at that facility, radiochemical analyses of tissues were initially performed at Battelle. However, in 1978 responsibility for radiochemical analysis of tissues was turned over to Los Alamos National Laboratory (LANL) under a separately administered program; the analyses on tissues originating at the Rocky Flats Facility continued there until the late 1980s, when funding and other considerations dictated their withdrawal from this activity.

Early efforts of the Transuranium Registry were directed toward identifying suitable populations of persons with occupational experience with plutonium and the higher actinides. Once these populations had been identified, workers were informed of the purpose of the Transuranium Registry and their voluntary participation as registrants was solicited. More than 1000 persons were ultimately registered (a number that has reduced over the years as additional knowledge and experience have been gained), and, by 1975, the results of 30 autopsies and tissue analyses were reported in the refereed literature (5). As of October 1, 1991, the Transuranium Registry had 467 living active registrants, including five whole-body donors with depositions estimated to be greater than 1.5 kBq, and had received tissues (autopsy or surgical specimens) from 265 donors, including nine whole-body donations.

A similar strategy of recruiting registrants was adopted by the Uranium Registry subsequent to its establishment but has not been nearly so successful. As of October 1, 1991, the Uranium Registry had 32 living registrants and had received tissues from one surgical case and 12 postmortem donors, including one whole-body donor. The total cohort of registrants, summarized by individual registry and birth decade, is presented in Table I.

The Registries: 1992 as a Year of Change

In February 1992, the DOE awarded a 3-year grant for $3.76 million to Washington State University (WSU) for management and operation of the registries. This was an important step in the continuing evo-

Table I. Birth Cohort Distribution of Living
Registrants

Birth Decade	Total	USTR	USUR
1900–1909	13	10	3
1910–1919	97	89	8
1920–1929	192	178	14
1930–1939	85	84	1
1940–1949	32	32	0
1950–1959	7	7	0
Birthdate unknown	73	67	6
Totals	499	467	3

lution of the registries and brought with it significant changes. The grant calls for management and operation of the registries as a single entity rather than as parallel but administratively separate research programs. In addition, the radiochemistry support now provided by the LANL under separate contract to the DOE will be carried out under subcontract with WSU beginning with the second year of the grant, thereby providing fully integrated management for the entire program.

Combined management and operation of the registries and the radiochemistry operations at LANL should not only provide for better integration and centralized control of the programs but should also reduce overhead and direct operating costs. One obvious benefit is the elimination of redundant efforts and forms arising from the existence of two separate registry entities.

In addition, other significant advantages include enhanced opportunities for collaboration with the broad spectrum of faculty available at a major research university. In particular, WSU offers some unique opportunities in this regard through its Health Research and Education Center (HREC), which, along with the registries, is a part of the College of Pharmacy. The HREC was created by the Washington State legislature in 1989 to carry out research in biomedical and social health. Medical support for the registries is provided through the HREC, which, in conjunction with the WSU Electron Microscopy Center, offers unique opportunities for histopathology studies utilizing the registries' collection of microscopic pathology materials. Other opportunities include the specialized analytical capabilities, including use of an inductively coupled plasma mass spectrometer and a 1-MW TRIGA-fueled reactor for neutron activation analysis.

Perhaps the most innovative and potentially advantageous aspect of the transfer of the program to WSU is the integration of the registries into the academic programs of the university, thereby providing a mechanism for training students in health physics and radiobiology, two disciplines historically in short supply of practitioners. Conventionally, support for students in health physics has been via a grant or fellowship directly to the student, normally (but not always, depending on the fellowship) with an equal amount provided to the institution. The DOE fellowship program is by far the largest and supports about two dozen graduate students annually. Each receives $15,000 per year for support, and the institution receives a similar amount. Typically, only one or two fellowship students attend any given institution, and the institutional grant, although generous on a per student basis, is insufficient to support even a single faculty member, let alone an entire program.

By providing direct faculty research support, as is the case with the registries' grant, a critical mass of faculty can be readily achieved and the opportunity for student thesis research is created. Hence, students can be more readily informed of, and attracted to, these disciplines. In addition, the grant provides support (including tuition) for two half-time student research assistants; the actual annual outlay for each of these students is about $15,000 or half of what the cost would be if the conventional fellowship mechanism were used. And, as an added bonus, these two students provide invaluable assistance in furthering the research carried out by the registries.

Finally, two additional benefits attributable to the location of the program at a major university should be stressed: enhanced credibility and academic freedom. In recent years government and government-funded programs have been subjected to increasing scrutiny from the public as well as from their peers. Not surprisingly, considering the nature of the research, which involves postmortem collection and analysis of tissues from workers known to have had intakes of plutonium or other actinides, the registries have not been immune from such scrutiny and inquiries by the media. Therefore, it is essential to ensure that the scientists performing the work are unfettered by the funding agency in the scientific conduct of the program and that the research be carried out in an open and ethical fashion. As the DOE has recognized, this condition is best accomplished through a grant to an independent and recognized research university.

Administratively, the registries are centered on the Tri-Cities campus of WSU, and specialized laboratory and medical support staff are located in Spokane. The registries' staff includes three full-time faculty members—two radiobiologists and a health physicist who serves as director. This nucleus of researchers is supported by two half-time graduate student research assistants and a full-time administrative assistant. Medical support is provided by the Director of the WSU Health Research and Education Center, a full-time faculty member who devotes a portion of his time to serving as the registries' medical director. Radiochemistry support is provided by LANL under direct contract to the DOE and includes two professional radiochemists, with special expertise in actinide chemistry, and one technician. As noted previously, with the commencement of the second year of the grant in February 1993, the radiochemistry operations will be administered directly by the registries via a subcontract with LANL.

The grant also provides for the addition of a fourth faculty member and a third half-time student research assistant in the third year of the grant. Plans are also being considered to add another faculty member and half-time research assistant to manage and operate the National Human Radiobiology Tissue Repository.

Research Objectives of the Registries

The primary objective of the registries is to ensure the adequacy of radiation protection standards for the actinide elements, verifying or modifying, as appropriate, the existing biokinetic and dosimetry models on which the standards are based. This objective is accomplished by a carefully structured program of research designed to evaluate the distribution, concentration, and biokinetics of the actinide elements in humans. Tissues collected at autopsy from volunteer donors with a history of exposure to the actinides are radiochemically analyzed to determine their content of actinide nuclides. These results are evaluated, along with radiation exposure and medical histories, and compared with estimates of body, lung, and other organ burdens made during life with measured postmortem deposition to assess the validity of biokinetic and dosimetric models on which the standards are based and to develop refinements or modifications to these models based on actual human experience. In addition, the registries also compare the results of animal experiments with those obtained from the human tissue studies to gauge the validity of interspecies comparisons. Another important function is the evaluation of histopathology slides and other specific human data to assess toxic changes possibly attributable to actinide exposure and to provide basic data for the determination of risk coefficients for radiation exposure.

Finally, the registries act as a repository for information on internal deposition of actinides in humans and encourage and carry out collaborative research with other groups of scientists. During the 1991–1992 time frame, active collaborations were carried on with no less than 15 institutions (Figure 1). Collaboration with and direct assistance to other researchers will be facilitated through the creation of the National Human Tissue Archive for radiological specimens. In addition to solutions of tissues, histopathology slides and blocks, and remaining unanalyzed tissues from USTUR cases, this archive will contain tissues collected by Argonne National Laboratory for the Radium Dial Painter Study. This unique collection of human tissue materials, plus other donated tissues collected from people with histories of radioactivity intake, will thus be made available to scientists studying the effects of radioactivity in humans.

Operation of the Registries

The basic registries' operation can be described in terms of a five-step process. The first step, identification of potentially suitable donor populations or individuals, has historically been accomplished through contacts made via the employer of the potential registrant, because

- Argonne National Laboratory
  Surface Deposition of Actinide in
  Human Bone
  Oncogene Studies
- Georgetown University
  Postmortem External Radioactivity
  Measurement, Case 1001
- Inhalation Toxicology Research
  Institutute
  Autoradiography and Microscopic
  Examination of Respiratory Tract,
  Case 246
  Histopathology Study of Osteosarcoma,
  Case 262
- Lawrence Berkeley Laboratory
  Soft Tissue Autoradiography, Case 246
- Los Alamos National Laboratory
  Radioochemical Analysis of Tissues
  Numerous Special Projects and Studies
- National Cancer Institute
  Risk Estimates and Epidemiology of
  Thorotrast
  Evaluation of Case 1001
- National Institute of Standards and
  Technology
  Radiochemical Intercomparison Studies
  and Development of Standard
  Reference Material—Human Bone
- National Naval Medical Center
  Medical, Autopsy and Postmortem
  Radioactivity Measurements,
  Case 1001

- National Radiological Protection
  Board (Great Britain)
  Distribution of Actinide in Human Bone
  Autoradiography of Bone
- Pacific Northwest Laboratory
  Biokinetic Modeling of Uranium
  Actinide Distribution in the Human
  Skeleton
  Comparison of Skeletal Actinide
  Distribution in Humans & Animals
  Distribution of Actinide in the
  Respiratory Tract
  Postmortem Direct Radioactivity
  Measurements, Cases 246 and 1001
  Soft Tissue Autoradiography Studies
- Saint Mary's Hospital
  Data Base Automation, Uranium Miner
  Lung Cancer Study
- United Kingdom Occupational
  Radiation Exposure Study
  (UNIKORNES)
  Assistance with Establishment of British
  Registry
- University of California, Davis
  Scanning Bone Density Study
- University of Pittsburgh
  Distribution of Actinide in the
  Respiratory Tract
- University of Washington
  Diurnal Excretion of Uranium

Figure 1. Research institutions collaborating with USTUR, 1991–1992.

virtually all exposures of interest are incurred in the workplace. Whether done on a group or individual basis, as might be the case with an individual specifically identified by the plant medical or health physics staff as of potential interest to the registries, the mechanism is essentially the same. Through their employers, potential registrants are provided with general information about the registries and, if interested, invited to contact the registries directly, either by collect telephone call or postage-paid card.

The next step is the actual enrollment process. Once a positive expression of interest from an individual has been received by the registries, the purposes and operations of the registries are again explained orally and in more depth, and the individual is provided with a detailed written description of the program. If the potential registrant remains interested, specific information regarding his or her exposure history is sought to determine if he or she will make a suitable donor. Suitability is largely a matter of prior exposure history; acceptance criteria are based on a documented and confirmed deposition or intake of one or more actinide nuclides, typically at levels of a few tens of a becquerel (Bq) or greater. If the potential donor desires to become a registrant and is acceptable to the registries, formal voluntary donation and acceptance are accomplished through the completion of informed consent, permission for autopsy, and medical and health physics records release forms.

Registrants enrolled in the program are sent a brief letter annually to update them on the status of the registries and to request updated information regarding changes in address or employment. Autopsy permissions or whole-body donation forms are renewed on a 5-year cycle at which time new informed consent forms are also obtained. Each registrant is issued a personal dated identification card and, if desired, a Medic Alert registration and identification bracelet or necklace.

Registrants are enrolled as either routine autopsy or whole-body donors. Whole-body donors are volunteers with depositions typically exceeding 150 Bq and who have a well-documented exposure history or other characteristics that would make them of scientific interest. This requirement, along with a natural reticence to make a whole-body donation as opposed to an autopsy, severely limits the pool of potential whole-body donors, and most volunteers are therefore accepted as routine autopsy donors.

The next step in the process involves the actual collection of tissues. This collection is normally accomplished postmortem except for those few instances in which surgical specimens are collected or the individual is a participant in a special study that involves the collection of excreta or blood during life. The postmortem tissue collection protocol of the USTUR evolved on the basis of experience and availability of cases. Initially, samples were routinely obtained of lung, tracheobronchial lymph nodes, liver, and bone (1, 6, 7). After the first few autopsies, the collection protocol was expanded to include the entire liver and both lungs, plus samples from one or more of the following: thyroid, kidney, spleen, gonads, muscle, and fat. Further experience with the autopsy procedure and subsequent radioanalytical results led to the development of an expanded formal autopsy tissue collection protocol, as described by Breitenstein (6) and Kathren (7), which has been recently refined and is detailed in Figure 2.

The registries also request paraffin blocks or prepared histopathological slides of the various tissues collected. These are typically examined at autopsy by the private pathologist performing the autopsy. Slides and blocks are saved and made part of the National Human Tissue Archives.

Tissue

Lungs (entire, with associated nodes)
Lymph Nodes (Hilar)
Liver (whole or minimum of 400g)
Bone:
   Ribs (one or more, typically
      left 6 and 7 and
      excluding 1,2,11,12)
   Sternum (whole)
   Vertebral wedge (lumbar, 3 contiguous)
   Patella (both)
   Clavicle (one)
Spleen (whole)
Kidneys (both)
Ovaries or testes (both)
Prostate
Rectus muscle*
Body fat*
Stomach*
Esophagus*
Thyroid*
Heart*
Tumor *
Wound Site and Associated Nodes

*Sample: ≥ 20g

Figure 2. Routine autopsy tissue collection protocols of USTUR.

All tissues collected are subject to radiochemical analysis to determine their actinide content. These data are entered into a newly developed, computerized database and evaluated on an individual case basis, as well as collectively, along with relevant information relating to exposure, excretion, and bioassay data collected during life, medical history, and autopsy results to gain additional understanding of the distribution, biokinetics, and dosimetry of the actinides from actual human experience.

Toward Improved Biokinetic Models for Plutonium and Americium

To ensure the adequacy of radiation protection standards for the actinides and thus to achieve the basic goal of the registries, it is essential that the standards be based on sound biokinetic models. Accordingly, much of the research effort of the registries has been directed toward biokinetic or, as they were sometimes called in the past, metabolic models. A major step was taken with the evaluation of the first whole-body donation, which was published as a compendium of five articles constituting the entire October 1985 issue of *Health Physics* (8). This case, identified as USTUR Case 102, involved a chemist who had incurred an accidental deposition of ^{241}Am as a result of a wound some 25 years prior to death. At the time of death, his measured total body burden was 5.5 kBq (147.4 nCi) of ^{241}Am, of which more than 80% was resident in the skeleton and only about 7% in the liver (9). This distribution pattern differed significantly from that predicted by the then current model of the International Commission on Radiological Protection (ICRP), which predicted more nearly equal amounts in the skeleton and liver. (10) The postmortem radioassay data, along with bioassay and other health physics information obtained during life, were used to develop and evaluate a new five-compartment model for ^{241}Am based solely on human data (11). One of the key features of this model was a retention half-time of only 2–3 years for ^{241}Am in the liver in contrast to the then accepted value of 40 years based on analogy with Pu and animal data.

Further support for a retention half-time of 2–3 years for ^{241}Am in liver was obtained from a subsequent study of the relative distribution of ^{238}Pu, ^{239}Pu, and ^{241}Am in the skeleton and liver of occupationally exposed individuals, using tissues obtained at autopsy by the USTUR (12). This finding of a shorter effective clearance time for ^{241}Am in liver has significant implications for the dose delivered to the liver from a given intake of Am and hence the radiation protection standards for that nuclide.

Other recent and continuing work of the registries deals directly with the application and evaluation of the validity of existing biokinetic models (13–15). One such study compares estimates of systemic deposition made by six laboratories using urinalysis data on a series of 17 individuals with estimates made on the basis of postmortem radiochemical analysis of tissue (13). Typically, the estimates made by the six laboratories were in good agreement with each other but were consistently greater than the estimates made from postmortem tissue analysis. The deviation between the urinalysis and autopsy estimates appeared to be inversely related to the level of Pu in the body (i.e.,

the smaller the estimated deposition, the greater the ratio of the urinalysis to autopsy estimate with convergence of the two occurring at
estimated burdens of about 1 kBq).

Another recent study involving comparison of premortem and
postmortem estimates of plutonium in skeleton and liver was carried
out jointly by the registries and the Pacific Northwest Laboratory (14).
Skeletal and liver depositions of six former workers at the Hanford
site were evaluated using an empirically developed model for internal
use based on that of Jones (15) and ICRP Publication 48 (16). Organ
burdens estimated from urinary excretion data were found to be roughly
consistent with those made from postmortem tissue analysis. Individual estimates were within a factor of 3 for skeleton, a factor of 5 for
liver, and a factor of 2 for skeleton and liver combined. In general,
urinalysis estimates of skeletal deposition tended to be greater than
autopsy estimates, but the converse was true for the liver.

A more recent study compared estimates of plutonium deposition
calculated with various biokinetic models with actual measurements of
the plutonium content of the whole body after death (17). This comparison was done with five whole-body donations to the registries. The
urinary excretion data from these cases were used with several models
to obtain estimates of systemic deposition, and these results were
compared with the value measured in the tissues by postmortem radiochemical analysis. In general, the estimates made with the earlier
models were severalfold greater than the comparable postmortem
measured values and consistent with what would be expected on the
basis of the previous comparison study. Estimates made with more
recent models, such as those put forth by Jones (15), Leggett (18),
and Leggett and Eckerman (19), were generally in close agreement
with the measured postmortem values.

Recently, the registries utilized data from postmortem analysis of
whole-body donations to develop a new biokinetic model for [241]Am
(20). The new model can be compared with that put forth in ICRP
Publication 48, which is the generally accepted model (16). The ICRP
48 model assumes that once the [241]Am reaches the transfer compart-
ment—is absorbed—45% is deposited in the skeleton and 45% in the
liver with half-times of 50 and 20 years, respectively. The remaining
10% is characterized as going to early excretion. Thus, the fractional
long-term deposition, $R(t)$, at t years after intake can be characterized
by the following two-compartment exponential equation:

$$R(t) = 0.45e^{-0.014t} + 0.45e^{-0.035t} \qquad (1)$$

By contrast, the registries' model, based on actual human data (19), uses the parameters expressed in Table II and can be expressed in terms of a three-compartment exponential equation:

$$R(t) = 0.45e^{-0.014t} + 0.025e^{-0.28t} + 0.30e^{-0.069t} \tag{2}$$

Similarly, a new model can be developed from the whole-body data for plutonium as reported and compared with the ICRP Publication 48 model in current use (16). ICRP 48 uses the same biokinetic constants for both Am and Pu, and hence the mathematical representation is identical for both and is characterized by equation 1. Using the radiochemical data from five whole-body cases (21), along with health physics measurements and information of when the intakes may have occurred, the biokinetic parameters shown in Table III were developed for plutonium and lead to the mathematical representation shown in equation 3:

$$R(t) = 0.4e^{-0.014t} + 0.4e^{-0.035t} + 0.2e^{-0.069t} \tag{3}$$

The differences between the ICRP model for both Pu and Am, characterized by equation 1, and the registries' model (2) for Am (equation 2) and Pu (equation 3) are significant and should lead to refinement and improvement in the estimation of in vivo deposition and dose estimates.

Conclusions

The human tissue studies of the registries are important to understanding the mechanisms by which the actinide elements move

Table II. Biokinetic Parameters for [241]Am

Compartment	Fractional Uptake	Residence Half-Time (years)
Skeleton	0.45	50
Liver	0.25	2.5
Muscle	0.20	10
Rest of body	0.10	10

Table III. Biokinetic Parameters for [239]Pu

Compartment	Fractional Uptake	Residence Half-Time (years)
Skeleton	0.4	50
Liver	0.4	20
Muscle	0.2	10

throughout the body and are of potential immediate practical application to the safe use of uranium and the transuranium elements. Perhaps the most important application is the verification or indicated refinement of existing biokinetic models upon which internal dose calculations and radiation protection standards are based.

Another important practical application is the verification of operational health physics estimates of deposition made by in vivo counting or other bioassay techniques. Tissues from people with radioactivity uptakes are of enormous potential value in the study of oncogenes and biomarkers, as well as for more traditional studies of possible radiation-induced pathology. The increased understanding of the biokinetics, measurement, dosimetry, and biological effects of actinides in humans promised by the human tissue studies of the registries is essential to maintaining and ensuring a suitably safe workplace for those involved with the various actinide elements; no amount of animal data, circumstantial evidence, or calculation can ensure that the radiation protection standards applied to humans are, in fact, both safe and reasonable. We can only be certain that our understanding of the actinides in humans is correct if it has in fact been gained from the proper study and interpretation of actual human experience.

Note Added to Proof

Since the preparation of this chapter in early 1992, there have been numerous changes in the operations and activities of the registries. The National Human Radiobiology Tissue Repository became a reality in the latter half of 1992, with radiation biologist John J. Russell as its curator. The registries themselves are still part of the Washington State University College of Pharmacy but now are administratively a part of Health Physics and Radiobiology Research of the college. They have acquired a half-time faculty member, Scott E. Dietert, who serves as resident medical consultant. On February 1, 1994, the registries assumed complete responsibility for the performance of radiochemistry operations, which will be carried out by WSU under the direction of radiochemist Roy E. Filby and will include a provision for training graduate students in radiochemistry. After a brief overlap period, radiochemistry support from Los Alamos National Laboratory will be phased out completely by late 1994. The original two faculty members—health physicist Ronald L. Kathren, who serves as director, and radiobiologist Ronald E. Filipy—remain with the program. Lynn A. Harwick serves as administrative manager.

References

1. Nelson, I. C.; Heid, K. R.; Fuqua, P. A.; Mahoney, T. D. *Health Phys.* **1972**, *22*, 925–930.

2. Newton, C. E., Jr.; Heid, K. R.; Larson, H. V.; Nelson, I. C. *Tissue Sampling for Plutonium through an Autopsy Program*; AEC Research and Development Report BNWL-SA-918; Pacific Northwest Laboratory: Richland, WA, 1966.

3. Newton, C. E., Jr.; Larson, H. V.; Heid, K. R.; Nelson, I. C.; Fuqua, P. A.; Norwood, W. D.; Marks, S.; Mahoney, T. D. In *Diagnosis and Treatment of Deposited Radionuclides*; Kornberg, H. A.; Norwood, W. D., Eds.; Proc. 7th Ann. Hanford Biology Symp.; Excerpta Medica Foundation: Amsterdam, Netherlands, 1968; pp 460–468.

4. Bruner, H. D. In *Diagnosis and Treatment of Deposited Radionuclides*; Kornberg, H. A.; Norwood, W. D., Eds.; Proc. 7th Ann. Hanford Biology Symp.; Excerpta Medica Foundation: Amsterdam, Netherlands, 1968; pp 661–665.

5. Norwood, W. D.; Newton, C. E., Jr. *Health Phys.* **1975**, *28*, 669–675.

6. Breitenstein, B. D., Jr. In *Actinides in Man and Animals*; Wrenn, M. E., Ed.; RD Press: 1981; pp 269–272.

7. Kathren, R. L., Ed. *Radiat. Prot. Dosim.* **1989**, *26*, 323–330.

8. Roessler, G. R., Ed. *Health Phys.* **1985**, *49*(4), 559–661.

9. McInroy, J. F.; Boyd, H. A.; Eutsler, J.; Romero, D. *Health Phys.* **1985**, *49*, 587–621.

10. ICRP. *Limits for Intakes of Radionuclides by Workers*; ICRP Publication 30; *Annals of the ICRP* **1979**, *2*(3/4), 1–116.

11. Durbin, P. W.; Schmidt, C. T. *Health Phys.* **1985**, *49*, 623–661.

12. Kathren, R. L.; McInroy, J. F.; Reichert, M. M.; Swint, M. J. *Health Phys.* **1988**, *54*, 189–194.

13. Kathren, R. L.; Heid, K. R.; Swint, M. J. *Health Phys.* **1987**, *49*, 623–661.

14. Sula, M. J.; Kathren, R. L.; Bihl, D. E.; Carbaugh, E. H. In *Proc. 7th Int. Conf. of the International Radiation Protection Association*, Sydney, Australia, April 10–17, 1988.

15. Jones, S. R. *Radiat. Prot. Dosim.* **1985**, *11*, 19–27.

16. ICRP. *The Metabolism of Plutonium and Related Elements*; ICRP Publication 48; *Ann. ICRP* **1986**, *16*(2/3), 1–98.

17. Kathren, R. L.; Heid, K. R.; Swint, M. J. *Health Phys.* **1987**, *53*, 487–493.

18. Leggett, R. W. *Bioassay Data and a Retention-Excretion Model for Systemic Plutonium*; NUREG/CR-3346; ORNL/TM-8795, 1984.

19. Leggett, R. W.; Eckerman, K. F. *Health Phys.* **1987**, *52*, 337–346.

20. Kathren, R. L.; McInroy, J. F. *J. Radioanal. Nucl. Chem.* **1992**, *156*, 413–424.

21. McInroy, J. F.; Kathren, R. L. *Radiat. Prot. Dosim.* **1989**, *26*, 151–158.

RECEIVED for review August 7, 1992. ACCEPTED revised manuscript March 25, 1993.

Lung Cancer Mortality and Radon Exposure

A Test of the Linear–No-Threshold Model of Radiation Carcinogenesis

Bernard L. Cohen[1] and Graham A. Colditz[2]

[1]University of Pittsburgh, Pittsburgh, PA 15260
[2]Harvard University, Cambridge, MA 02138

The linear–no-threshold theory used to estimate the cancer risk of low-level radiation from the known risks of high-level radiation is tested by studying the variation of lung cancer mortality rates (m) with average exposure to radon (r) in various U.S. states and counties. The data indicate a strong tendency for m to decrease with increasing r, in sharp contrast to the theory prediction of a strong increase of m with increasing r. To explain this discrepancy by a strong tendency for areas of high radon to have low smoking prevalence, and vice versa, would require almost 100% negative correlation between radon and smoking, whereas current information indicates a correlation of only a few percent. Several other possible explanations for the discrepancy are explored, but none seem to be effective in substantially reducing it.

THE CANCER RISKS FROM LOW-LEVEL RADIATION are usually estimated by use of a linear theory, assuming that risk is proportional to exposure. Our purpose here is to test that theory by studying the relationship between lung cancer risk and exposure to radon in homes.

The lower part of Figure 1 shows plots of age-adjusted lung cancer mortality rates for males and females vs. average radon exposure in

0065–2393/95/0243–0067$08.00/0

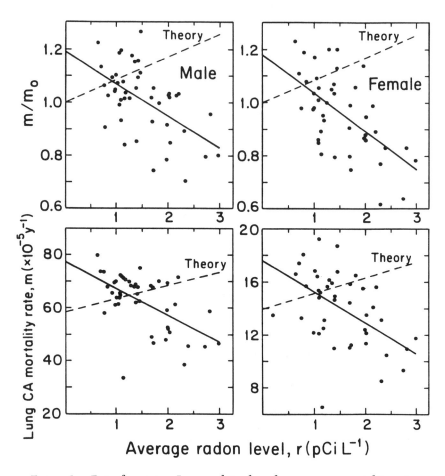

*Figure 1.   Data for states. Lower plots show lung cancer mortality rate,
m, vs. average radon level, r. Upper plots show m/m$_o$ vs. r in which
m$_o$ is the correction for smoking frequency (eq 5). Solid lines are the
best fit of the data to eq 5, and dashed lines are the predictions of
BEIR-IV theory; these predictions are calculated in the lower plots, with
m$_o$ for each state taken to be the national average.*

various U.S. states. The data suggest a negative slope, lung cancer
rates decreasing with increasing radon exposure, whereas the theory
predicts a positive slope (dashed line)—radon causes lung cancer, so
increasing radon exposure should increase lung cancer rates. We refer
to this difference between the expected positive slope and our observed negative slope as our *discrepancy*.

The lower part of Figure 2 shows similar data for more than 900
counties, except that rather than showing more than 900 data points,
we have divided the abscissa into intervals, as shown at the top mar-

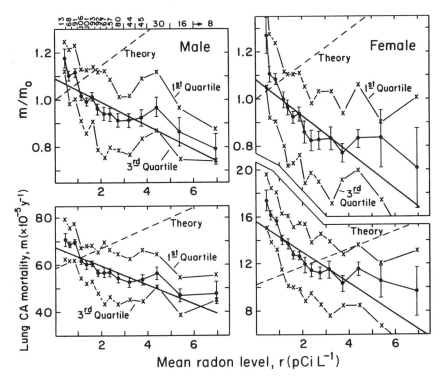

Figure 2. Data for counties. Similar to Figure 1, except that rather than showing points for separate counties, the abscissa is divided into intervals, and plots show only characteristics of the distribution of counties in the interval: the mean value of the ordinate and its standard deviation and the value of the ordinate for the first and third quartiles. Figures at the top are the number of counties in each interval.

gin, and have plotted the characteristics of the distribution of data points in each interval—its mean ordinate, the standard deviation (SD) of the mean, and the first and third quartiles. Again, the data indicate a strong negative slope, but the theory requires a strong positive slope.

This discrepancy with theory, although surprising, could be explained if there were a strong negative correlation between radon and smoking; that is, if areas with low radon had a greater population of smokers than areas with high radon. To develop a quantitative treatment of this explanation, we start at a basic level.

The National Academy of Sciences Report (*1*), known as BEIR-IV, predicts a lifetime mortality risk, m', to an individual due to his lifetime exposure to radon, r', as

$$m' = a(1 + br') \tag{1}$$

where $b = 10.8\%$ per picocurie per liter and a is the risk with no radon exposure, which varies by substantial factors for smokers and nonsmokers, both male and female. We then sum equation 1 over all the people in the county (or state) and divide by the population, P. The sum of the risks to all individuals divided by P is the mortality rate, m. The sum of the radon exposures to all individuals divided by P is the average radon exposure, r. If the fraction of the adult population that smokes is S, and hence the fraction that does not smoke is $(1 - S)$, our sum gives

$$m = [Sa_s + (1 - S)a_n] (1 + br) \tag{2}$$

where the subscripts refer to smokers (s) and nonsmokers (n). With a correction for migration (2)—the fact that people do not spend their entire lives in their county of residence at time of death—equation 2 becomes

$$m/m_o = 1 + Br \tag{3}$$

where m_o and B are mathematical expressions arising from the conversion of equation 2 to equation 3 with numerical values (for 1970 to 1979 mortality rates)

$$B = 0.073 \text{ for counties and } 0.083 \text{ for states} \tag{4}$$

$$m_o = 9 + 99S \text{ for males and } 3.7 + 32S \text{ for females} \tag{5}$$

Equation 3 provides the correction for smoking frequency we are seeking.

A Bureau of Census Survey (3) contains reasonably good data on smoking frequency for states. These are used to calculate m_o from equation 5, and the upper parts of Figure 1 are plots of m/m_o vs. r. The discrepancy between theory and observation remains. In the lower part of Figure 1, the discrepancies in the slopes are 6.8 SD for males and 5.7 SD for females, and in the upper part, the discrepancies are 7.3 and 7.1 SD, respectively.

The reason the correction for smoking does not resolve the discrepancy is that a very low correlation exists between radon levels and smoking frequency. The correlation (R^2) is only 6.7% for males and 0.7% for females.

The Bureau of Census Survey (3) gives our best estimates of S (we refer to these as our "preferred" S values), but a completely independent set of data comes from cigarette sales tax collection in various years (4), which we take to be an indicator of S for males. The

r–S correlations are 1960—6.4%, 1970—2.1%, and 1975—1.3%, and the discrepancies between theory and observations are larger than for our preferred S values. Another source of data on S was tried, and it gave an r–S correlation of 1.8%.

We had no direct data on S for counties, but a crude approximation would be to assume it to be the same as for the states. A better approximation is to apply a correction for the well-known rural–urban difference in smoking frequency. When values of S thus derived are used to calculate m_o, the dependence of m/m_o on r is as shown in the upper parts of Figure 2. Again, the discrepancies in the slopes B are little affected—22 SD for males and 19 SD for females in the lower part of Figure 2 vs. 21 and 17 SD, respectively, in the upper part. The r–S correlation is only 4.3% for males and 4.5% for females.

However, perhaps our data on S are erroneous, and the r–S correlation really is strongly negative. How strong a correlation would be necessary to explain our discrepancy? We worked this out by using a model and found that elimination of the negative slope observed in Figure 2 would require a 65% r–S correlation, but even a perfect r–S correlation would not give a positive slope as strong as that predicted by the theory.

Even if we had no information on S, how likely is an r–S correlation of 65% or higher? The fraction S is, to a large extent, a socioeconomic variable (SEV), and we have data for each country on many SEVs. We considered 37 of these variables* and determined their correlations with r. The three largest correlations were 17.7, 12.4, and 7.4%; 5 SEVs had correlations between 5.5 and 4.4%; and none of the other 29 SEVs had a correlation larger than 3%. This finding makes the 4.3% correlation between r and S in our data seem quite reasonable, and it would appear to be essentially incredible for the errors in our data to be large enough and systematic enough to give an r–S correlation of 65% or higher.

We conclude that problems with our data on S cannot explain our discrepancy.

*Population: total, percent increase from 1970 to 1980, per square mile, males/females; Income: median household, per capita, average wage for manufacturing workers, dollars per capita retail sales, percent below poverty level, percent unemployed, percent with more than one car; Age: median, percent over 65, percent of births to mothers younger than 20; Education: percent high school graduates, percent college graduates, dollars per capita for education; Housing: percent owner occupied, percent under 10 years old, percent over 50 years old, median value, building permits/100 units, average persons per household; Medical: physicians per capita, hospital beds per capita, percent of budget for health; Urbanization: percent living in urban areas, percent of labor force in manufacturing, percent of land in farms, farm earnings as percent of total; Social: crime rate, percent white, divorce rate; Government: dollars per capita local taxes, percent of budget for police, percent of budget for welfare, percent of vote to leading party (1984).

Perspective on Our Discrepancy

Some perspective on our discrepancy can be obtained by studying the relationship between radon and smoking for various types of cancer. We studied those types of cancer that have been tentatively linked to radon in published papers (5). Because there are no theories of the details of this linkage, we used double regression, fitting the data to

$$m = e + fr + gS \qquad (6)$$

where e, f, and g are parameters adjusted to fit the data. The results are listed in Table I including the t ratios, $t(r)$ and $t(S)$, the number of SD by which f and g in equation 6 differ from zero, and R^2, the percentage of the variation of m that is explained by equation 6.

Table I shows that not only is $t(S)$ large and positive for lung cancer, as expected, but $t(r)$ also is several times larger for lung cancer than for any other cancer type. The effect of omitting the S dependence in equation 6 is shown in the last two lines of Table I in which R^2 is still much higher for lung cancer than for other cancer types. Clearly, the relationship between radon and lung cancer is special; this result is expected, but the problem is that the sign of the relationship is negative rather than positive. Our discrepancy is a unique phenomenon, with nothing comparable to it in other types of cancer.

Table I. Results for Various Cancer Types

Cancer Type	Sex	$t(r)$	$t(S)$	R^2
Lung	M	−8.5	13.4	26.0
Lung	F	−8.9	10.2	20.0
Leukemia	M	2.2	1.0	0.6
Leukemia	F	0.3	0.8	0.1
Melanoma	M	−2.4	4.7	3.6
Melanoma	F	−1.8	0.2	0.4
Kidney	M	−0.2	−1.5	0.2
Kidney	F	2.4	1.0	0.7
Prostate	M	−0.1	−3.5	1.4
Lymphoma	M	2.0	−0.8	0.6
Lymphoma	F	2.6	0.2	0.8
Lung	M	−10.6	—	11.0
Lung	F	−10.8	—	11.0

NOTE: These are the results of fitting our data for counties to

$$m = c_0 + c_1 r + c_2 S$$

The values of $t(r)$ and $t(S)$ represent the number of SDs by which c_1 and c_2, respectively, differ from zero; R^2 is the percent of the variations of m that is explained by this equation. The last two rows are the results with c_2 set equal to zero.

Is Our Discrepancy Due to Coincidences in Our Data Set?

Possibly, the negative slopes B are a property of our particular data set, arising from some unrecognized coincidence. The simplest way to test for this possibility is to divide our data into subsets in various ways and to analyze each subset independently. This step was done by stratifying on an individual SEV and dividing the data set into quintiles. For example, in stratifying on population (P), the first quintile (Q-1) consists of the 20% of our counties with the lowest population (most rural), and the fifth quintile (Q-5) consists of the 20% with the highest population (very urban). The data for each of these subsets are then fit with equation 3 to derive the value of the best-fit slope B. We stratified in this way on each of our 37 SEVs in turn to obtain (5 quintiles × 2 sexes × 37 SEVs =) 370 different data subsets, giving 370 B values. Of these 370 B values, 369 are negative and the single exception is easily explainable as a statistical fluctuation. The average of these B values is very close to the values from the total data set: -0.047 vs. -0.050 for males and -0.072 vs. -0.077 for females.

Thus, the phenomenon of large negative B values applies separately and independently if we consider only the very rural or very urban counties, if we consider only the richest or poorest, if we consider only the fastest growing or those with declining population, if we consider only the most educated or least educated, if we consider only those with the best health care or those with the poorest health care, and so on, and it also applies to all the strata in between. It clearly is not caused by some unusual coincidence.

As further evidence of this matter, we have analyzed data collected by the Environmental Protection Agency in 22 states and found a negative slope similar to the slope we found for our data in those states. A similar negative slope also was reported (6) for the counties in England and Wales.

Confounding Factors

As in any epidemiological study, results can be influenced by confounding factors (CFs) that correlate strongly but for unrelated reasons with both m and r and thereby introduce an apparent correlation between m and r that is not caused by a direct cause–effect relationship. Smoking prevalence is ab initio the best candidate for a CF because of its known strong correlation with m, but we have investigated its effects earlier. Most other potential CFs that we can imagine would correlate with SEVs. For example, air pollution might be a CF, and it correlates with several of our SEVs, like population density and percent urban. These SEVs then act as surrogates for the CF and can be substituted for it in mathematical analyses. We therefore consider each of our SEVs to be a potential CF.

Stratifying our data on a CF would greatly reduce the problem of confounding, as each data subset (i.e., each stratum) would have quite similar values for the CF. The average slope B obtained from the five quintiles should then approximate the correct value, free of the confounding effect.

The stratification studies described in the previous section include tests of this process for each of our 37 SEVs. In no case is the average slope B substantially different than the value of B derived from the entire data set. If our problem is a CF, none of our 37 SEVs comes close to serving as a surrogate for it.

This still leaves the possibility that several of our CFs combine to give a large effect. The best available method for studying this is through multiple regression, assuming that

$$m/m_o = 1 + Br + c_1X_1 + c_2X_2 + \ldots + c_{37}X_{37} \tag{7}$$

where X_1, X_2, $\ldots$, are our 37 SEVs and c_1, c_2, $\ldots$, are adjustable parameters selected, along with B, to give the best fit to the data.

Fitting our data with equation 7 rather than with equation 3 reduces the derived value of B from -0.050 to -0.015 ± 0.005 for males and from -0.077 to -0.027 ± 0.010 for females, discrepancies with theory of 16 and 10 SD, respectively. These values appear to substantially reduce our discrepancy.

However, the literature on multiple regression is full of warnings against the foregoing procedures and rarely are more than four or five variables considered appropriate in seeking causal relationships. The reason for this caution is easily understood. Because r is correlated with m, any new variable that is correlated with m will be somewhat correlated with r and therefore drain away some of the dependence of m on r in finding the best fit to equation 7.

We investigated this effect with a model in which the SEVs were constructed as a linear combination of m and a random number, with relative weights selected to give the same correlation with m as our actual SEV. They were not constructed to have any correlation with r, so they are not CFs. Using these constructed SEVs, we obtain values of B very similar to those obtained with the actual SEV. This result indicates that the reduction in our discrepancy by use of multiple regression analyses, using equation 7 rather than equation 3, is largely due to the mathematics of multiple regression rather than to the true effects of confounding.

Stratification on Geography

Geography is the only factor known to correlate strongly with r; therefore, we stratified on it. The U.S Bureau of Census divides the nation into four regions, with each region consisting of two or three divisions. The results of treating each of these as a separate data set are listed in Table II.

Table II shows that for each of the four national regions, the slope B is negative by more than 2 SD for both males and females, and the average values of B for the four regions are not significantly different from the values obtained for the nation as a whole, -0.042 vs. -0.050 for males and -0.072 vs. -0.077 for females. Stratifying on geography to the level of national regions does very little to reduce our discrepancy.

However, Table II shows that stratifying further to the level of divisions does have an appreciable effect; 5 of the 18 B values are positive, and the average values of B are substantially reduced, to -0.023 for males and -0.053 for females. This result reduces the discrepancy with the prediction of BEIR-IV theory, $B = +0.073$, by 22 and 16%, respectively, from the discrepancy without stratification.

This finding suggests that finer stratification on geography might be useful. For 18 states, our data file contains mean radon levels in 20 or more counties. We treat the counties in each of these states as

Table II. Regional and Divisional Results

Region —Division	Number of Counties	Male B	Male t	Female B	Female t
Northeast	202	−0.050	−4.6	−0.087	−5.4
—New England	63	+0.016	+0.5	+0.045	+0.9
—Mid Atlantic	139	−0.055	−4.3	−0.113	−6.4
North Central	358	−0.019	−2.3	−0.030	−2.6
—East NC	196	+0.012	+1.1	+0.017	+1.1
—West NC	162	−0.015	−1.3	−0.034	−2.1
South	235	−0.047	−3.1	−0.095	−4.4
—South Atlantic	155	−0.030	−1.2	−0.068	−2.0
—East S. Central	54	−0.042	−2.7	−0.071	−2.5
—West S. Central	26	−0.131	−3.5	−0.198	−3.7
West	116	−0.051	−2.4	−0.074	−2.1
—Mountain	97	−0.020	−0.8	−0.055	−1.3
—Pacific[a]	19	+0.051	+1.2	−0.003	−0.0
Averages					
Regions	228	−0.042	−3.1	−0.072	−3.6
Divisions	101	−0.023	−1.2	−0.053	−1.8

[a]Pacific includes only Washington and Oregon.

a separate data set, and the results of analyzing them are listed in Table III. Nine of the 36 B values are positive and the average B values are slightly less negative than those for the divisions, $B = -0.015$ vs. -0.023 for males and -0.051 vs. -0.053 for females. On the whole, finer stratification on geography from divisions to individual states does relatively little to reduce our discrepancy, but it does achieve the maximum reduction of the discrepancy we have found, 28% for males and 17% for females.

Stratifying on geography per se introduces important risks of confounding problems. For example, an ethnic group that is unusually susceptible to lung cancer may happen to live in a high radon area, a situation leading to a positive value of B. On a national scale, effects of such chance correlations would strongly tend to average out, but in a limited area, they could be very important.

An obvious problem with stratifying on geography is poor statistics. Four of our nine geographic divisions have fewer than 64 data points, and 12 or our 18 individual states provide fewer than 40 data points. With a small number of data points, a fit to a line of various slopes can more easily occur by chance.

The results in Tables II and III indicate that geography is probably a reasonably important confounding factor, but the negative slopes B and the large discrepancy with theory still remain. On the other

Table III. Results for Individual States

State	Number	Male		Female	
		B	t	B	t
CO	25	−0.041	−0.8	−0.072	−0.9
GA	20	−0.102	−1.0	−0.290	−2.4
ID	39	−0.003	−0.1	−0.120	−1.4
IL	37	+0.004	+0.2	−0.058	−1.7
IN	33	+0.013	+0.6	−0.001	−0.03
IA	88	−0.021	−1.3	−0.012	−0.5
MD	22	−0.072	−1.6	−0.098	−2.1
MI	35	+0.050	+2.0	+0.029	+1.5
MN	41	−0.015	−1.1	−0.021	−0.9
NJ	21	−0.009	−0.2	−0.0004	−0.01
NY	55	+0.010	+0.4	−0.043	−1.2
NC	34	−0.024	−0.6	+0.002	+0.1
OH	52	−0.001	+0.1	+0.010	+0.5
PA	63	−0.008	−0.5	−0.054	−2.4
TN	29	−0.002	−0.1	−0.007	−0.1
VA	46	+0.018	+0.4	−0.123	−1.5
WV	22	−0.022	−0.5	−0.072	−1.3
WI	39	−0.055	−1.6	+0.008	+0.1
Average		−0.015	−0.32	−0.051	−0.79

hand, these results suggest the desirability of obtaining much more data to do a better job of stratifying on geography.

Other Linear–No-Threshold Theories

All of the preceding treatments are based on the BEIR-IV theory. Several other linear–no-threshold theories have been proposed, differing principally in their treatment of smoking, which is not well established from the data on miners. Other parameters, based on total risk of lung cancer and increased risk to miners with high radon exposure, are subject to much less uncertainty and therefore to much less variation among different theories.

We have shown that the discrepancies described previously with the BEIR-IV theory apply equally to all other theories. The reasons for this are easily understood. Smoking is essentially not correlated with radon exposure; therefore, the treatment of smoking makes no difference; based on the miner data, all theories predict a similar strong positive slope for m vs. r, whereas the data in Figures 1 and 2 clearly show a strong negative slope.

References

1. U.S. National Academy of Sciences Committee on Biological Effects of Ionizing Radiation. *Health Risks of Radon and Other Internally Deposited Alpha Emitters (BEIR-IV);* National Academy Press: Washington, DC, 1988.
2. Cohen, B. L. *Int. J. Epidemiol.* **1990,** *19,* 680–684.
3. *Smoking and Health;* Office of Smoking and Health, U.S. Public Health Service: Rockville, MD, 1990; DHHS Pub. No. (CDC) 87-8396 (Revised 02/90).
4. *The Tax Burden on Tobacco;* Tobacco Institute: Washington, DC, 1988.
5. Henshaw, D. L.; Eatough, J. P.; Richardson, R. B. *Lancet* **1990,** *335,* 1008.
6. Haynes, R. M. *Radiat. Prot. Dosim.* **1988,** *25,* 93–96.

RECEIVED for review March 7, 1992. ACCEPTED revised manuscript March 25, 1993.

Evidence of Cancer Risk from Experimental Animal Radon Studies

Fredrick T. Cross

Pacific Northwest Laboratory, Box 999, Richland, WA 99352

Epidemiologic data from underground miners confirm that radon decay products are carcinogenic, but evidence for the quantitative risks of these exposures, especially for indoor air, is less conclusive. Experimental animal studies, in conjunction with dosimetric modeling and molecular–cellular level studies, are particularly valuable for understanding the carcinogenicity of human radon exposures and the modifying effects of exposure rate, the physical characteristics of the inhaled decay products, and associated exposures to such agents as cigarette smoke. Similarities in animal and human data, including comparable lung cancer risk coefficients, tumor-related dosimetry, and tumor pathology, presently outweigh their differences. The animal models, therefore, appear to be reasonable substitutes for studying the health effects of human radon exposures.

IN THE 1940S IT WAS BELIEVED THAT THE RISKS of exposure to radon were potentially important only to underground miners extracting ores containing radium and uranium (1). Since that time occupational studies of uranium and other underground miners have yielded consistent estimates of the lung cancer risk associated with exposure to radon (2–4). That evidence has also been substantiated by studies of animals exposed to radon (5). The potential hazards of indoor radon exposure, essentially unrecognized in the 1940s, have been studied only com-

0065–2393/95/0243–0079$08.00/0

paratively recently (6), particularly since the discovery in 1984 of a house in Pennsylvania containing radon concentrations several thousand times greater than levels in most houses. Even more recent is the presumed association of radon exposure and cancers of organs other than the lung (7, 8).

Studies of radon-induced lung cancer in experimental animals are particularly valuable for understanding the carcinogenicity of human radon exposures in the home and workplace. Animals can be exposed to a variety of agents under carefully controlled conditions and then sacrificed for the study of developing lesions or observed throughout their life span for tumor development. The doses to critical cells in the respiratory tract can be determined, and these in turn can be related to doses to critical cells in the respiratory tract of humans exposed to similar aerosols.

The study of radon-induced mutations, changes in expression of oncogenes and tumor suppressor genes, and growth factors and growth factor receptors during tumor progression in animals also provides valuable evidence on the underlying mechanisms of radon carcinogenesis. This evidence, particularly that of the efficiency for oncogenic transformation at low dose rates, is crucial to the determination of the risk of lung cancer from exposure to indoor levels of radon.

This chapter reviews the evidence for radon-induced cancer in experimental animals and emphasizes the carcinogenicity of radon exposures in rats. The few mechanistic data on radon-induced lung tumors in rats currently available are not reviewed here.

Health Effects Data

Radon health effects data, developed primarily in adult male animals, are provided by the Pacific Northwest Laboratory (PNL) and the Compagnie Générale des Matières Nucléaires (COGEMA) laboratory in France (5). Approximately 800 Syrian Golden hamsters, 6000 SPF Wistar rats, and 100 beagle dogs were exposed to mixtures of radon, radon progeny, diesel engine exhaust, uranium ore dust, and cigarette smoke in PNL studies; about 10,000 SPF Sprague–Dawley rats were exposed to mixtures of radon, radon progeny, ambient (outdoor) aerosols, and cigarette smoke in COGEMA laboratory studies. Additional French radon carcinogenesis modeling studies have employed intramuscular injections of the promoter 5, 6-benzoflavone to further clarify the role of promoters in radon-induced cancers (9). The rat data from the two laboratories are discussed as a whole, primarily because of their similarity; emphasis, however, is placed on the PNL data. Data from other animal species, discussed only briefly here, were presented in greater detail in the report to the U.S. Department of Energy (5).

Major biological effects produced in the radon studies were respiratory tract tumors [adenomas, bronchioloalveolar (BA) carcinomas or adenocarcinomas, epidermoid carcinomas, adenosquamous carcinomas, and sarcomas], pulmonary fibrosis, pulmonary emphysema, and life-span shortening (5). Appreciable fibrosis, emphysema, and life-span shortening, although somewhat species dependent, did not occur at exposure levels less than 3.5 J h m^{-3} [1000 working-level month (WLM); working level (WL) is defined as any combination of short-lived radon decay products in 1 L of air resulting in the ultimate emission of 1.3 × 10^5 MeV of potential alpha energy (1 WL = 2.08 × 10^{-5} J m^{-3}). Working-level month is defined as an exposure equivalent to 170 h at 1 WLM concentration (1 WLM = 3.5 × 10^{-3} J h m^{-3})]. However, excess respiratory tract tumors were produced in rats at exposures considerably less than 0.35 J h m^{-3} (100 WLM), even at levels comparable to typical life-span exposures in homes (20 WLM). Further, tumors were produced in exposures to radon decay products alone; thus, associated exposures to other irritants, such as uranium ore dust or cigarette smoke, are not necessary for carcinoma development. With a few exceptions, the incidence of adenomas and sarcomas (both rarely found in control animals) was considerably less than 10%.

A decrease in exposure rate at a given exposure level not only increased the overall incidence of lung tumors but specifically increased the incidence of epidermoid carcinomas; a similar finding was noted in studies of the Colorado Plateau miners with protracted exposures (10). Protraction of exposures in rats also produced a significantly higher incidence of multiple primary lung tumors (more often of a different rather than the same type) and fatal primary lung tumors (11). Most (>70%) epidermoid carcinomas but only about 20% of adenocarcinomas were classified as fatal. Finally, most (−80%) radon-induced lung tumors in rats are considered to originate peripherally and to occur at the bronchiolar–alveolar junction, in contrast to human lung tumors, which generally are more centrally located. The remaining 20% of rat lung tumors are considered to be centrally located (bronchi associated); the actual percentage depends on exposure rate and possibly exposure level (11).

With the exception of the greater prevalence of solid alveolar tumors and bronchioloalveolar carcinomas [and the absence of small cell (K-cell) carcinomas] observed in rats, the evidence on cancer in rats and humans is reasonably consistent. Regional differences in sites of tumor formation are explained, in part, by dosimetry modeling (12). The doses to rat distal bronchioles and alveoli are generally quite high in comparison to doses to these locations in humans (miners). On the other hand, doses to miner proximal bronchi are generally quite high compared with those in the rat; thus, one might postulate that regions of tumor development coincide with regions of high dose and high sensitivity. Although the rat does not develop small cell carcinoma per se in response to radon exposures, the bombesin staining is similar in both rat epidermoid and human small cell carcinomas. This similarity suggests exploration of growth factor and growth factor receptor involvement in human and animal radiation-induced tumors.

Extrapulmonary lesions, including tumors, were produced primarily in the nose, particularly with high unattached fractions of radon decay products. Significant excess nonrespiratory neoplasms associated with radon exposure were previously noted primarily in the kidneys; however, recent data from the COGEMA laboratory show significant increases in bone, liver, and soft tissue cancers (13). The increase in bone sarcoma and liver cancer was noted at very low exposures comparable to lifetime exposures in most homes. The implications for human exposure are uncertain and will not be known until the susceptibilities, biokinetics, and dosimetry are compared across species. In exposures of COGEMA laboratory female Sprague–Dawley rats to 1600 WLM, the incidence of breast cancer doubled despite a significant reduction in life span attributable to mammary tumors. Again, the implication for human exposures is unclear. The scientists at PNL have exposed female Wistar rats but have not yet examined the resulting histopathology.

Other experiments at PNL have been performed to determine if prenatal effects could be produced by prolonged inhalation exposures to high concentrations of radon and radon decay products throughout gestation (14). Neither teratological nor reproductive effects were produced when pregnant SPF Sprague–Dawley rats were exposed to radon-progeny levels about 10,000 times the typical annual levels in houses. Thus, the human fetus is not expected to suffer teratological effects from typical indoor radon levels.

Factors Influencing Risk

The major factors found to influence the tumorigenic potential of radon exposures in laboratory rats include radon-progeny cumulative exposure, exposure rate, and unattached fraction (radon progeny not attached to airborne dust); associated cigarette-smoke exposures; and "time-since-exposure" (15). Respiratory tract cancer risk increases as radon-progeny cumulative exposure and unattached fraction increase and, as discussed previously, decreases with increase in radon-progeny exposure rate. Details of the cumulative exposure and exposure rate data are presented in the following section on risk modeling. The increased risk with high unattached radon progeny is particularly relevant to indoor radon exposures, where the unattached levels are generally much higher than those in underground mines. The PNL animal data project an approximate twofold increase in risk per WLM exposure for the typically five- to tenfold higher levels of unattached radon progeny in homes compared to mines.

The influence of associated cigarette-smoke exposures depends, in part, on the temporal sequence of radon-progeny and cigarette-smoke exposures. In the COGEMA laboratory experiments, the risk was synergistically increased when smoke exposures followed completed radon exposures, but the risk remained unchanged from radon-only exposures when the sequence of mixed exposures was reversed (16). The promotional effect of cigarette smoke was also seen for the preneoplastic lesion adenomatosis but not for lung tumors in recent PNL serial-sacrifice initiation–promotion–initiation (IPI) studies (17). Although analysis of the life-span IPI tumor data is not complete, current evidence suggests antagonism. Earlier PNL dog experiments (18) and recent mouse experiments at Harwell Laboratory (United Kingdom) (19) also showed antagonism in tumor production with alpha-particle radiation and cigarette-smoke exposures, possibly as a result of overly high radiation doses that obscured the promotional effect of cigarette smoke.

Considering the composite data, it now appears that radon and cigarette-smoke exposures are synergistic only under certain conditions of exposure. Preneoplastic lesions induced by radon exposure are promoted by cigarette smoking, but the incidence of tumors may not be increased if the exposure to cigarette smoke is not sufficiently prolonged. It is becoming increasingly clear that the duration of cigarette smoking is at least as important as, if not more important than, the number of cigarettes smoked daily. An earlier article by Doll and Peto (20) regarding British doctors who smoked presented the same conclusion, but this conclusion was not shared by other modelers of the data (21).

The time-since-exposure effect in radon carcinogenesis is also discussed in the following section.

Risk Modeling of Animal Data

Quantitative modeling of data from animal studies supplies risk coefficients that can be compared with similarly derived coefficients from epidemiologic data. Statistical analyses of lung tumor data from rats have been used to model the hazard using the Weibull function for the baseline risk. These baseline, age-specific risks, which have been estimated for experimental rats, are uncertain, in contrast to human lung cancer rates, which have been more carefully determined (22, 23). Figure 1 summarizes the results of analyses of PNL data based

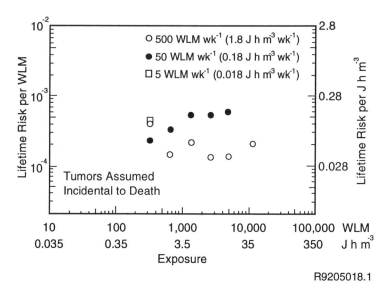

Figure 1. Lifetime risk coefficients for radon-progeny exposure of rats. (Adapted with permission from reference 22. Copyright 1989 NCRP Publications.)

on the linear relative risk model; in these analyses lung tumors are considered incidental to the death of the animal. As seen in the analysis of the COGEMA laboratory data, there is little indication of a decrease in risk per unit exposure with increasing total exposure, even to very high exposure levels. The analyses clearly show the influence on risk of exposure rate; the corresponding exposure-rate data are significantly different from each other, differing by factors of 2 to 3, with the exception of exposures at 1.1 J h m^{-3} (320 WLM).

The estimated linear-lifetime lung tumor risk coefficient, based on the combined exposure-rate data, was about 0.086 per J h m^{-3} (300 tumors per 10^6 rats per WLM) for adenomas and carcinomas combined. Excluding adenomas, the risk is reduced to about 0.071 per J h m^{-3} (250 cancers per 10^6 rats per WLM). These values may be compared to the overall (smokers and nonsmokers) National Research Council's Committee IV on the Biological Effects of Ionizing Radiations (BEIR IV) value of 0.10 per J h m^{-3} (350 cancers per 10^6 persons per WLM) and 0.040 per J h m^{-3} (140 cancers per 10^6 persons per WLM) for nonsmoking males (22). Estimates based on studies of male rats, therefore, are comparable to those obtained from human studies. Analyses based on the assumption that tumors are fatal produce risk coefficients about half as large. The lowest exposure-rate data [0.018 J h m^{-3} week^{-1} (5 WLM week^{-1})] in Figure 1 suggest that the exposure-rate effect (but not the risk) tapers off at lower exposure levels;

this effect will be tested in future epidemiologic analyses of exposures less than 1.1 J h m^{-3} (320 WLM).

The effects of exposure rate and time-since-exposure in the PNL experiments cannot be entirely separated. Figure 2 shows the risk versus age at which exposure stopped for the three exposure-rate groups. Although the pattern is not entirely consistent, the largest risks occurred in groups where exposure was protracted to older ages. The data in rats, therefore, appear to parallel the time-since-exposure effect observed in epidemiologic analyses of underground miners (4).

Even though there are differences in risks observed in rats with high exposure rates [1.8 J h m^{-3} week^{-1} (500 WLM week^{-1})] compared with those observed at lower exposure rates [0.18 J h m^{-3} week^{-1} (50 WLM week^{-1})], the implications for risks at typical residential exposure rates [$\sim$1.8 $\times$ 10^{-5} J h m^{-3} week^{-1} ($\sim$5 $\times$ 10^{-3} WLM week^{-1})] are not known and cannot be directly tested in a short-lived species such as the rat. The lowest exposure rate studied is somewhat comparable with those in former underground miners.

The two-mutation (recessive oncogenesis) model of Moolgavkar and Knudson (24) was tested with a PNL tumor data set similar to that used in the statistical analyses by Gilbert (22). This carcinogenesis model postulates transitions from a normal to an intermediate to a malignant cell with quantifiable transition rates and takes into account the growth characteristics of the normal and intermediate cell populations. The model describes the rat lung cancer data well (25). Briefly, the find-

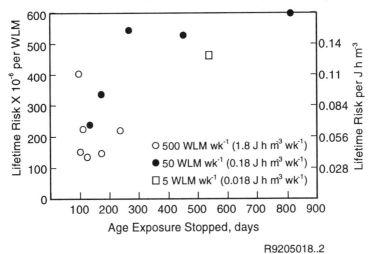

R9205018..2

Figure 2.   Lifetime risk in rats vs. age radon-progeny exposure stopped. (Adapted with permission from reference 22. Copyright 1989 NCRP Publications.)

ings are that the first mutation rate is very strongly dependent on the rate of exposure to radon progeny and the second mutation rate is much less so, suggesting that the nature of the two mutational events is different. The model predicts the following:

1. Radon doubles the background rate of the first mutation at an exposure rate of approximately 0.005 J h m^{-3} week^{-1} (1.35 WLM week^{-1}), an exposure rate definitely in the range of miner exposures.

2. Radon doubles the background rate of the second mutation at an exposure rate of about 1.4 J h m^{-3} week^{-1} (400 WLM week^{-1}) (consequently, the hypothesis that radon has no effect on the second mutation rate cannot be rejected).

3. The net rate of intermediate cell growth is doubled at a radon exposure rate of about 0.12 J h m^{-3} week^{-1} (35 WLM week^{-1}).

4. A drop occurs in hazard after radon exposures cease, paralleling the exposure-rate or time-since-exposure effect noted in the statistical risk analyses.

5. There is an optimal exposure schedule for producing tumors. Fractionation of exposure is more efficient in producing tumors, but further fractionation leads to a decreased efficiency of tumor production.

The implications of these findings for human risk assessment are also unclear at this time.

Discussion and Conclusions

A broad multilevel approach to radon cancer risk assessment includes mechanistic, animal, dosimetric, statistical, and carcinogenesis modeling data to infer risks to humans exposed in occupational and residential settings. The similarity of current adult rat and underground miner exposure-response data suggests that the rat model is particularly valuable for reducing scientific uncertainties in the human database, particularly in regard to the complex interactions of radon and cigarette-smoke exposures and the risks associated with childhood exposures. The effort to measure radon levels in schools in the United States demonstrates the concern for the latter type of exposures, although the observed drop in hazard with time-since-exposure would tend to discount early (e.g., childhood) exposures. The rat model is also valuable for delineating the mechanisms of radon carcinogenesis, as evidenced by recent studies on oncogene and growth factor–receptor involvement in radon-induced lung tumors in rats (*26, 27*).

Observations made in animal systems that have not been unequivocally found in human exposures to radon are (1) the increase in tumor production with increase in radon-progeny unattached fraction, (2) the importance of the temporal sequence of exposures to cigarette smoke and radon progeny, and (3) the occurrence of extrapulmonary and extrathoracic (head and neck) carcinomas. However, few data are available from epidemiologic studies on these aspects of the health effects of radon.

Acknowledgment

This work was supported by the U.S. Department of Energy under Contract No. DE–AC06–76RLO 1830.

References

1. Lorenz, E. *J. Natl. Cancer Inst.* **1944**, 5, 1–15.
2. National Council on Radiation Protection and Measurements. *Evaluation of Occupational and Environmental Exposures to Radon and Radon Daughters in the United States*; NCRP Report No. 78; NCRP Publications: Bethesda, MD, 1984.
3. International Commission on Radiological Protection. *Lung Cancer Risk from Indoor Exposures to Radon Daughters*; ICRP Publication 50; Pergamon Press: Oxford, United Kingdom, 1987.
4. National Academy of Sciences Committee on the Biological Effects of Ionizing Radiation. *Health Effects of Radon and Other Internally Deposited Alpha-Emitters (BEIR IV)*; National Academy Press: Washington, DC, 1988.
5. Cross, F. T. *Radon Inhalation Studies in Animals*; DOE-ER-0396; National Technical Information Service: Springfield, VA, 1988.
6. U.S. Department of Energy/Commission of European Communities. *International Workshop on Residential Radon Epidemiology*; CONF-8907178; National Technical Information Service: Springfield, VA, 1989.
7. Alexander, F. E.; McKinney, P. A.; Cartwright, R. A. *Lancet* **1990**, 2 June, 1336–1337.
8. Henshaw, D. L.; Eatough, J. P.; Richardon, R. B. In *Proc. 29th Hanford Symp. on Health and Environ.*; Battelle Press: Columbus, OH, 1992; pp 935–958.
9. Poncy, J. L.; Larouque, P.; Fritsch, P.; Monchaux, G.; Chameaud, J.; Masse, R. In *Proc. 29th Hanford Symp. on Health and Environ.*; Battelle Press: Columbus, OH, 1992; pp 803–819.
10. Saccomanno, G.; Archer, V. E.; Auerbach, O.; Kuschner, M.; Sanders, R. P.; Klein, M. G. *Cancer* **1971**, 27, 515–523.
11. Dagle, G. E.; Cross, F. T.; Gies, R. A. In *Proc. 29th Hanford Symp. on Health and Environ.*; Battelle Press: Columbus, OH, 1992; pp 659–675.

12. Cross, F. T. In *Proc. Workshop on the Future of Human Radiation Research*; BIR Report 22; British Institute of Radiology: London, 1991; pp 27–35.
13. Masse, R. In *Proc. 5th Int. Symp. on Natural Radiat. Environ.*, *Radiat. Prot. Dosim.* **1992**, *45*, 603.
14. Sikov, M. R.; Cross, F. T.; Mast, T. J.; Palmer, H. E.; James, A. C. In *Proc. 29th Hanford Symp. on Health and Environ.*; Battelle Press: Columbus, OH, 1992; pp 677–691.
15. Cross, F. T. In *Proc. 5th Int. Symp. on Natural Radiat. Environ.*, *Radiat. Prot. Dosim.* **1992**, *45*, 629.
16. Chameaud, J.; Perraud, R.; Chretien, J.; Masse, R.; Lafuma, J. In *Proc. 19th Hanford Life Sciences Symp.*; CONF-791002; National Technical Information Service: Springfield, VA, 1980; pp 51–57.
17. Cross, F. T.; Dagle, G. E.; Gies, R. A.; Smith, L. G.; Buschbom, R. L. In *Proc. 29th Hanford Symp. on Health and Environ.*; Battelle Press: Columbus, OH, 1992; pp 821–844.
18. Cross, F. T.; Palmer, R. F.; Filipy, R. E.; Dagle, G. E.; Stuart, B. O. *Health Phys.* **1982**, *42*, 33–52.
19. Priest, N. D.; Moores, S. R.; Black, A.; Talbot, R.; Morgan, A. In *Proc. 4th Int. Symp. Radiat. Prot.—Theory and Practice*; IOP Publishing Ltd.: Bristol, United Kingdom, 1989; pp 433–436.
20. Doll, R.; Peto, R. *J. Epidemiol. Community Health* **1978**, *32*, 303–313.
21. Moolgavkar, S. H.; Dewanji, A.; Leubeck, G. *J. Natl. Cancer Inst.* **1989**, *81*, 415–420.
22. Gilbert, E. S. In *Proc. 24th Mtg. Natl. Council on Radiat. Prot. Measurements*; NCRP Publications: Washington, DC, 1989; pp 141–145.
23. Gray, R. G.; Lafuma, J.; Parish, S. E.; Peto, R. In *Proc. 22nd Hanford Life Sciences Symp.*; CONF-830951; National Technical Information Service: Springfield, VA, 1983; pp 592–607.
24. Moolgavkar, S. H.; Knudson, Jr., A. G. *J. Natl. Cancer Inst.* **1981**, *66*, 1037–1052.
25. Moolgavkar, S. H.; Cross, F. T.; Leubeck, G.; Dagle, G. E. *Radiat. Res.* **1990**, *121*, 28–37.
26. Foreman, M. E.; McCoy, L. S.; Frazier, M. E. In *Proc. 29th Hanford Symposium on Health and Environ.*; Battelle Press: Columbus, OH, 1992; pp 649–655.
27. Leung, F. C.; Dagle, G. E.; Cross, F. T. In *Proc. 29th Hanford Symposium on Health and Environ.*; Battelle Press: Columbus, OH, 1992; pp 615–625.

RECEIVED for review August 7, 1992. ACCEPTED revised manuscript April 29, 1993.

Evaluating the Safety of Irradiated Foods

George H. Pauli

Division of Food and Color Additives, Center for Food Safety and Applied Nutrition, U.S. Food and Drug Administration, Washington, DC 20204

Health agencies throughout the world have evaluated the safety of irradiated foods by considering the likelihood that irradiation would induce radioactivity, produce toxic radiolytic products, destroy nutrients, or change the microbiological profile of organisms in the food. After years of study, researchers have concluded that foods irradiated under the proper conditions will not produce adverse health effects when consumed.

SAFE FOODS ARE IMPORTANT FOR ALL OF US. An adult eats approximately 1000 kg of food (solid and liquid) per year. Considering the total quantity, even small concentrations of harmful components are significant. Food safety is a many-faceted issue, however, and a variety of ideas exist regarding what is most important to ensure the safety of foods. Surveys often show a wide disparity in opinion between consumers and food scientists on what is most important. Generally, food scientists list overall dietary considerations (such as total calories or fat intake) and food-borne pathogens as much more important health factors than food processing, pesticides, or food additives, which are the primary concerns of consumers.

In addition, the concept of food safety can vary. For example, the safety standard is less strict when applied to inherent food components than when applied to added components. In the United States, the legal requirements for establishing safety or hazard depend on the situation. This variation greatly affects what is meant by safety. For example, the law allows foods to be sold as long as the natural components that might be toxic are not present in such large amounts as to make that food ordinarily injurious [Federal Food, Drug, and Cosmetic Act, Section 402(a)(1)]. Sale of a food may be prohibited, however, if added substances, such as environmental contaminants, are present in amounts whereby the food may be injurious [Federal Food, Drug, and Cosmetic Act, Section 402 (a)(1)]. Intentional use of substances, including radiation sources, that are reasonably expected to become components of food or to affect the characteristics of the food is prohibited until such use is demonstrated to be safe and a regulation is issued authorizing it [Federal Food, Drug, and Cosmetic Act, Sections 402(a)(2)(c) and 402(a)(7)]. Although the purpose of the law in each case is to ensure safe foods, the safety standard differs according to the situation.

Because there are several standards for safety, we must explicitly state which standard we are using when we discuss food safety. If something has caused observable harm, it is easy to conclude that it is unsafe. But it is impossible to prove the absolute absence of any potential for harm. Many natural components of food can be harmful under some circumstances. One can always hypothesize scenarios for harm that have not been, or cannot be, proven false with absolute certainty. If one were to insist on an absolute standard that excludes all possibilities for harm, one would have to conclude that nothing is safe and the word safety would become meaningless. Generally speaking, however, food safety scientists focus on whether there is any reasonable basis for presuming that a food is less safe than the food it is replacing in the diet.

Since 1958, irradiated foods may not be sold in the United States unless their safety has been demonstrated and regulations have been issued prescribing safe conditions of irradiation. The safety standard to be applied is a reasonable certainty that no harm will result from consuming such foods. Congress was explicit in recognizing that absolute certainty is an impossible, and thus meaningless, standard. The safety concerns for irradiated foods are the same concerns that apply to all foods. Food scientists have reached consensus that four areas need to be addressed: radiological safety, toxicological safety, microbiological safety, and nutritional adequacy.

Many studies throughout the world were begun in the 1950s to address these safety concerns. In the United States, the Department of the Army and the Atomic Energy Commission took the lead in sponsoring research. Internationally, committees of experts were established by the Food and Agricultural Organization (FAO) of the United Nations in Rome, the International Atomic Energy Agency (IAEA) in Vienna, and the World Health Organization (WHO) in Geneva. These

Joint (FAO–IAEA–WHO) Expert Committees on Food Irradiation (JECFI) met in 1964, 1969, 1976, and 1980 to provide guidance for research and to evaluate results. A unified worldwide program, the International Project in the Field of Food Irradiation, which began in 1970, was sponsored by the IAEA, the FAO, and the Organization for Economic Cooperation and Development in Paris. The WHO participated as a consultant. Eventually the resources of 24 countries were pooled for this program.

In this chapter I will only briefly discuss the resolution of the four safety concerns, summarizing what various expert groups have concluded. Several books comprehensively discuss the various technical studies done on irradiated foods (1–9). One book presents an excellent evaluation of the safety issues (9).

Radiological Safety

In the early days of research, radiation sources were still being developed. Fuel rods from nuclear reactors, containing a variety of radionuclides, were sometimes used. High-energy photons (or thermal neutrons) from some of these radionuclides had the potential to induce radioactivity in foods. Also, linear accelerators were used to produce electron-beam irradiation. Although electrons from machine sources have limited penetration capability, this capability can be increased by raising the voltage.

Research showed that electron-beam energies above 12 MeV induced detectable radioactivity in foods. Researchers rapidly settled on ^{60}Co and ^{137}Cs as safe gamma sources whose photon energies are too low to induce radioactivity. Energies of electron beams were restricted to 10 MeV. Later, recommendations of 5 MeV maximum voltage were established for X-ray sources (10). Countries permitting irradiation of food adopted these restrictions to ensure that radioactivity would not be induced in foods. Radiological safety of foods has not been an issue of concern with food scientists since these limitations were accepted.

Toxicological Safety

Toxicological safety testing often means animal testing, although this is but one aspect of toxicological safety evaluation. By using a complex, integrated biological system to represent the human response to a substance or mixture of substances, animal testing provides a comprehensive assessment of the many factors (dose, metabolism, and competing biochemical reactions) that combine to cause toxicity. Such a comprehensive testing approach, if sufficiently sensitive, may identify problems that could not otherwise be predicted.

Animal Studies. *Test Design.* Unless complemented by other types of safety testing, however, appropriate animal studies are difficult to design. In particular, careful attention must be given to whether the substances to be tested differ sufficiently from the control substances for toxicity to be assessed in the animals. In other words, money and time spent for animal toxicity testing are wasted unless the effects of irradiation on food are so large that the radiolytic products formed have a reasonable potential to produce a toxic effect when the food is fed in the largest amount compatible with good nutrition.

Moreover, animal testing of whole foods can be misleading because small differences in animal health between treated and control groups, which can occur by random statistical variation, can be confused with adverse effects caused by treatment. Normally, whether such differences are random or caused by treatment can be evaluated by noting whether consumption of larger amounts of the substance causes an increase in the effect. However, this cannot be done with foods because of physical and nutritional limits on the amounts that can be consumed.

Effect of Dosage. Toxicological testing is usually conducted by administering the test substance in amounts far in excess of what humans would consume. This excess dosage is intended to provide a safety factor (or uncertainty factor) to compensate for insufficient knowledge about which species best represents humans and for insufficient statistical power to detect small increases in the incidence of adverse effects. However, attempts to provide a safety factor by use of excess dosage can pose problems. For example, increasing the dose of radiation may result in a test with a food that is inedible, or increasing the amount of an irradiated food in the diet may result in a diet that does not meet basic nutritional needs. Many tests have been done under conditions that were, in retrospect, unrealistic for assessing safety (*11*). Thus, any evaluation of safety that uses animal tests must be consistent with basic principles of safety evaluation, which include the following points:

- The dose makes the poison.
- If an effect is real and of general importance, it will be reproducible under a variety of circumstances.
- If no adverse effects are seen under severe test conditions (e.g., high dose or continuous exposure), they are very unlikely to occur under milder conditions. However, effects seen under severe conditions may or may not reflect what happens under milder conditions, although they do indicate a need for caution.

- The effort spent to test for safety should be commensurate with the potential risk.

Nutritional Requirements.   To draw valid conclusions from animal testing, one must thoroughly know and understand the test systems used, the chemical composition of the substances tested, and the needs and susceptibilities of the test animals. In the early days of irradiated food testing, when relatively little was known about the chemical effects of irradiating foods or of the nutritional needs of animals, appropriate feeding tests were difficult to design. The likelihood of eliciting toxic effects from feeding irradiated food depends on the concentrations and toxic potential of the radiolytic products in the food. With little knowledge about the identities or concentrations of radiolytic products, and without a large historical control database gained from years of animal studies, it was difficult to determine whether an animal feeding study would be helpful in assessing the safety of the food.

Likewise, an attempt to compensate for these difficulties by feeding large amounts of a particular irradiated food often led to nutritional problems such as unbalanced diets, which adversely affected treated and control groups, or diets of marginal nutritional adequacy, which resulted in nutritionally significant differences between treatment and control groups even though a relatively small amount of nutrients was lost because of irradiation. As a result, many of the early animal-feeding studies raised more questions than they answered and led to stricter requirements for permitting the sale of irradiated foods.

Chemical Effects of Radiation on Food.   Until more could
be learned about the chemical effects of irradiation and about the cause of adverse effects in these early studies, attempts were made to improve study designs, test more foods, and use more experimental animals. In the United States, lifetime feeding studies in rats and mice, chronic feeding studies in dogs, and reproduction studies in both of these species were conducted.

Toxicological Safety.   Increased knowledge of radiation chemistry and animal nutrition, however, contributed greatly to our understanding of toxicological safety. A better understanding of nutrition showed that the adverse effects occasionally reported in animal studies may be caused by nutritional deficiencies resulting from inappropriate test design. Food irradiation studies may have contributed more to the knowledge of animal nutrition than vice versa (12).

In this regard, toxicity and nutrition issues must be separated. In toxicity studies, the animal is fed the test substance in the largest

amount it can tolerate so that subtle effects can be discerned. However, when food is used as the test substance, other foods are dis-
placed, and nutrient deficiencies can result. Similarly, if the radiation
dose is increased, the amount of radiolytic products and, thus, the
sensitivity of a toxicity test are likewise increased. However, nutrient
loss is dose-dependent. Therefore, unless the diet contains an excess
of nutrients, an increased radiation dose can result in nutritional deficiencies.

Test Design. A greater understanding of radiation chemistry allowed researchers to design tests that can resolve specific questions,
interpret specific tests, and reduce speculation. Several examples illustrate the usefulness of this knowledge.

Within the dose range applicable to food irradiation, the quantity
of radiolytic products is proportional to the irradiation dose. The identity of radiolytic products does not change appreciably if the food and
radiation conditions remain constant. These results may seem obvious,
but until reliable data were obtained, there was sometimes a reluc-
tance to extrapolate from one dose to another or to evaluate one test
in light of other tests in similar foods.

Direct and Indirect Effects. The chemical effects of radiation on
food can be either direct or indirect. The food molecule that absorbs
the energy can react directly, or free radicals formed by the absorbing
species (such as water) can diffuse through the food and react with
other molecules. The relative proportion of these two types of reactions can vary with physical conditions. For example, in rigid matrices
such as dry foods or frozen foods, indirect effects become minor and
the overall chemical effect is substantially reduced.

Indirect effects, thus, depend not only on the radiant energy but
also on diffusion properties and the reactivities and concentrations of
various components in the food. Therefore natural food components,
such as antioxidants, can inhibit some types of reactions. In addition,
although research with model systems may be useful for understanding the process, oversimplified model systems may not always represent the response of foods to irradiation. For example, the effects
of radiation on food components in dilute aqueous solutions are much
greater for each radiation dose than on the same components in complex matrices. Because nearly all the absorbed energy is eventually
transferred to the diluted solute, the effective dose for the solute is
much greater than the dose for the solution as a whole.

Early genetic toxicity testing with simple models such as irradiated
sugar solutions showed reproducible mutagenic effects in vitro that
were not duplicated by tests in more complex systems (*13*). Appar-

ently either the mutagens are formed in much smaller concentrations in complex food matrices or they decompose more quickly. Similarly, radiation effects on nutrients are much larger in simple solutions than in whole foods.

Food Safety Assessment. So far, it may not be apparent how this chemical knowledge has enhanced our ability to assess safety. After all, considering the enormous chemical complexity of food, one can never understand all the chemical reactions that take place when food is irradiated. It is also impossible to identify all the radiolytic products and measure their toxicities. A wide variety of chemical reactions results from the immense number of chemical components in food. Irradiation of them will create an enormous number of radiolytic products at extremely low concentrations. For this reason, it is as impractical to assess the safety of an irradiated food on the basis of its components as it is to assess the safety of that food before irradiation based on its individual natural components. However, one can compare irradiated and nonirradiated foods to assess the significance of any changes. Two different approaches to this assessment were used by the U.S. Food and Drug Administration (FDA) and by several other countries.

FDA Toxicological Principles. In the early 1980s, the FDA published the toxicological principles that it applies to food safety (*14*). It is impossible to know everything about safety, so the FDA concentrates its efforts on those issues with the greatest risk potential. Because radiation can affect a whole food and thus a substantial portion of the diet, irradiation could pose a significant risk if its effect were large or unusual. Current knowledge of radiation chemistry, however, allows estimation of how great an effect irradiation will have on food. The effect is directly proportional to dose, all other things being equal. The wide diversity of radiation-induced reactions and the complexity of food ensure that individual radiolytic products will be formed in extremely low concentrations.

Bureau of Foods Irradiated Food Committee. An FDA committee, the Bureau of Foods Irradiated Food Committee (BFIFC), was assembled in 1979 to recommend appropriate criteria for evaluating the toxicological safety of irradiated foods (*15*). BFIFC found that radiolytic products formed in foods were essentially the same as substances found in other foods that were not irradiated. Any differences were not remarkable; specific compounds formed as a result of irradiation but not found in other foods were essentially similar chemically to other food components.

However, minor constituents of foods are not well characterized at the low concentrations comparable to those of radiolytic products. Therefore, a comparison of the concentrations of substances formed by radiation with those of the substances present in other common foods will be incomplete. Thus, as a caution to prevent underestimation of potential for risk, this committee assumed that 10% of the concentration of radiolytic products formed would be substances not otherwise consumed in food. These products were called unique radiolytic products because they may not be present in food that is not irradiated, although there was no evidence that they could be unique in any other regard. This name has caused some confusion, because some people have misinterpreted it as meaning unique to nature or unique in regard to potential toxicity.

On the basis of its experience with toxicity testing and its observation from chemical analyses, BFIFC concluded that any properly conducted animal-feeding test (of any duration) of a food irradiated at a dose below 1 kilogray (kGy) would not show a toxic effect. Recognizing that this is equivalent to saying that such foods are safe to eat, the committee recommended that foods irradiated at such doses be considered toxicologically safe without explicit confirmatory animal testing. A corresponding recommendation was issued for minor food ingredients irradiated at doses as high as could be foreseen (50 kGy), because the consumption of such foods was sufficiently small to make any risk from radiolytic products negligible. BFIFC also recommended that safety decisions on major food components irradiated at a dose above 1 kGy be based on 90-day animal-feeding studies and a genetic toxicity screen, with further testing if necessary to clarify inconclusive results from basic screening tests. These recommendations served as the basis for FDA decisions in the 1980s.

Joint Expert Committees on Food Irradiation. The JECFI was also making recommendations to the Codex Alimentarius Commission, a United Nations-sponsored organization established to promote harmonization of food regulations and thereby encourage world trade. In 1976 it recommended that several irradiated foods be considered safe (*16*). In 1980 JECFI concluded that the irradiation of any food commodity up to an overall average dose of 10 kGy presents no toxicological hazard (*17*). This conclusion was based on the chemiclearance principle, integrating the results from animal-feeding and in vitro studies with what is known about the chemical effects of irradiation. This conclusion, often supplemented by evaluations of national committees, today serves as the basis for irradiated food laws in many nations. For example, the United Kingdom Advisory Committee on Irradiated and Novel Foods (ACINF) concluded that there is no evidence to suggest that any toxicological hazard to human health would arise from food irradiated up to an overall average dose of 10 kGy (*18*).

Animal-Feeding Studies. In the United States an FDA task force reviewed all animal-feeding and in vitro studies for which it could obtain data, a total of several hundred studies. Because of the large volume of data, the FDA established a screening procedure to identify those animal-feeding studies that met 1980 criteria for design, conduct, and reporting. These studies were then used for in-depth review. In addition, all studies that reported adverse effects were evaluated carefully, even those that were deficient in some regard. Therefore, a study that reported adverse effects was not overlooked simply because it did not meet all of the standards for testing. Finally, the task force looked for trends in reported effects to determine whether any alleged but unproven adverse effects were confirmed by further testing.

The task force concluded that none of the studies showed adverse toxicological effects attributable to irradiated food (*19*). For the reasons discussed earlier, some of the studies reported adverse effects that appeared, on further consideration, to be nutritional effects caused by improper diet. The task force reported that its findings were consistent with BFIFC's conclusion that toxic effects should not be observed from foods irradiated at a dose below 1 kGy and concurred that such foods are safe. It also concluded, however, that few irradiated-food animal-feeding studies met all 1980 test standards.

It recommended that safety decisions for foods irradiated at higher doses and consumed in significant amounts be made on a case-by-case basis, after considering data specific to the situation. Since the task force review, FDA has completed an evaluation of animal-feeding studies using irradiated chicken. On the basis of these data, the FDA concluded that no toxicological hazard exists from poultry irradiated at a dose below 3 kGy.

Microbiological Safety

Food poisoning usually refers to food contaminated by microorganisms or by toxins from microorganisms. Indeed, microbiological safety is the most important safety concern for most foods. Although one would expect irradiation, which kills microorganisms, to improve the microbiological safety of food, this safety concern also needs careful evaluation.

Processing Method. As with other food-processing methods, the effects of irradiation vary with the type of food and microorganism.

Some microorganisms are very susceptible to irradiation, whereas others are resistant. With any food-processing method, one must ensure that reduction or elimination of one organism does not promote the growth of other organisms of greater health concern. This could occur either through elimination of competition or by apparent extension of shelf life so that pathogens could grow without apparent signs of spoilage.

We are familiar with these problems in the commercial sterilization of heat-treated food. It must be heated sufficiently, without recontamination, to eliminate the most heat-resistant of pathogens, *Clostridium botulinum*. Pasteurized food is not treated so severely but requires continued refrigeration. The same issues arise with irradiated food.

Mutations in Organisms. A second issue common to irradiation and thermal processing is whether irradiation could cause mutations in organisms, making them pathogenic or more virulent. Although mutations can be caused by irradiation or heat, such mutations generally are not beneficial to the organism and there is no evidence of any problem caused by irradiation-induced mutations. FDA concluded in 1986 that there is no reason to expect that mutants resulting from irradiation would be any different or more virulent than those created in nature (*19*).

Surviving Pathogens. With regard to pathogens that survive radiation treatment, JECFI concluded in 1980 that the microbiological safety achieved by irradiation is fully comparable to that of other currently accepted treatments (*17*). ACINF concluded in 1986 that "although irradiation up to an overall average dose of 10 kGy would not kill all pathogenic microorganisms, and could allow continued growth of surviving pathogens, the same possibilities arise with all of the accepted non-sterilising methods of food processing . . ." (*18*).

The FDA requires evidence that irradiation under prescribed conditions will not prevent spoilage that would allow a resistant pathogen to grow undetected to hazardous levels (*20*). This requirement is part of establishing good manufacturing practices that are necessary for any food-preservation method. The FDA also recognized that an irradiation dose below 1 kGy will destroy few spoilage bacteria in food and thus will not change normal spoilage patterns (*21*).

Nutritional Adequacy

Nutritional issues apply not so much to individual foods as to the overall diet. Thus nutrient losses, which occur in all food processing, must be understood in the context of the diet. With irradiation, the only nutrient losses of concern are of micronutrients, such as vitamins. The amount of macronutrients, such as protein, in a food is so large that the amount lost by irradiation is negligible.

Processing Conditions. Nutrient losses can also be mitigated by processing conditions. For example, because freezing minimizes indirect chemical effects, fewer nutrients are lost by frozen meats irradiated at very high doses in the absence of air than by meats irradiated at lower doses without such precautions (17). Similarly, because of the radiation protection provided by some food components, the effects of irradiation on a specific vitamin may vary with the individual food.

One needs to assess the likelihood that irradiation could affect the nutritional quality of the diet rather than simply measuring nutritional losses in individual foods. A relatively large loss of a nutrient from a food that is not a significant source of that nutrient may be of little importance. A smaller relative loss from a food that is an important source of a nutrient results in a greater absolute effect.

Evaluation of Nutritional Loss. At this time, all evaluations of specific irradiated foods have concluded that nutrient losses caused by irradiation are insignificant to human nutrition. In 1980 JECFI concluded that the effect of irradiation on the nutritional value of food should be compared with other processes, and a considerable body of data gives no cause for concern (17). In 1984 the FDA concluded that available data demonstrate that food irradiated up to 1 kGy has the same nutritional value as a comparable food that has not been irradiated (21). In 1986 ACINF concluded that irradiation at the appropriate dose, up to an overall average dose of 10 kGy, does not have any special adverse effect on the nutritional content of food. It recommended, however, that use of irradiation processing and any nutritional consequences in the light of consumption patterns should be monitored (18).

Summary

A large body of physical, chemical, microbiological, toxicological, and nutritional data on irradiated foods was generated over a 40-year period and reviewed by health agencies worldwide. On the basis of these data and the advice of scientific experts, the Codex Alimentarius Commission adopted the Recommended International Standard for Irradiated Foods, which recommends allowing foods to be irradiated up to an overall average dose of 10 kGy. Governments of individual countries took different regulatory approaches. Some adopted the Codex Standard in toto, whereas others saw no need for the technology. In general, even those countries that express reservations about the potential misuse of the technology have not disagreed in principle on the safety of food treated in accord with adequate standards (22). The FDA position is that there are no safety concerns at a dose below 1 kGy and that safe conditions for irradiation at higher doses should be established on a case-by-case basis.

References

1. *Preservation of Food by Ionizing Radiation*; Josephson, E. S.; Peterson, M. S., Eds.; CRC Press: Boca Raton, FL, 1982; Vol. I.
2. *Preservation of Food by Ionizing Radiation*; Josephson, E. S.; Peterson, M. S., Eds.; CRC Press: Boca Raton, FL, 1983; Vol. II.
3. *Preservation of Food by Ionizing Radiation*; Josephson, E. S.; Peterson, M. S., Eds.; CRC Press: Boca Raton, FL, 1983; Vol. III.
4. *Food Irradiation*; World Health Organization: Geneva, Switzerland, 1984.
5. *Radiation Preservation of Foods*; Advances in Chemistry Series No. 65; American Chemical Society: Washington, DC, 1967.
6. *Radiation Chemistry of Major Food Components: Its Relevance to the Assessment of the Wholesomeness of Irradiated Foods*; Elias, P. S.; Cohen, A. J., Eds.; Elsevier: Amsterdam, Netherlands, 1977.
7. *Recent Advances in Food Irradiation*; Elias, P. S.; Cohen, A. J., Eds.; Elsevier: Amsterdam, Netherlands, 1983.
8. Urbain, W. M. *Food Irradiation*; Academic Press: Orlando, FL, 1986.
9. Diehl, J. F. *Safety of Irradiated Foods*; Dekker: New York, 1990.
10. *Codex Alimentarius Volume XV:* Codex General Standards for Irradiated Foods and Recommended International Code of Practice for the Operation of Radiation Facilities Used for the Treatment of Foods. Food and Agriculture Organization of the United Nations and World Health Organization: Rome, Italy, 1984.
11. Kraybill, H. F.; Whitehair, L. A. *Annu. Rev. Pharm.* **1967,** *7*, 357–380.
12. *Symposium on New Aspects of Nutrition Uncovered in Studies with Irradiated Foods, Fed. Proc.* **1960,** *19*, 1023.
13. Beyers, M.; Den Drijver, L.; Holzapfel, C. W.; Niemand, J. G.; Pretorius, I.; Van der Linde, H. J. In *Recent Advances in Food Irradiation;* Elias, P. S.; Cohen, A. J., Eds.; Elsevier Biomedical: Amsterdam, Netherlands, 1983; pp 171–188.
14. *Toxicological Principles for the Safety Assessment of Direct Food Additives and Color Additives Used in Food;* U.S. Food and Drug Administration; Goverment Printing Office: Washington, DC, 1982; p 10.
15. Brunetti, A. P.; Fratalli, V.; Greear, W. B.; Hattan, D. G.; Takeguchi, C. A.; Valcovic, L. R. *Recommendations for Evaluating the Safety of Irradiated Foods*; U.S. Food and Drug Administration; Government Printing Office: Washington, DC, 1980.
16. *Wholesomeness of Irradiated Food:* Report of the Joint FAO/IAEA/WHO Expert Committee. World Health Organization Technical Report Series 604; World Health Organization: Geneva, Switzerland, 1977.

17. *Wholesomeness of Irradiated Food*: Report of a Joint FAO/IAEA/WHO Expert Committee; World Health Organization Technical Support Series 659; World Health Organization: Geneva, Switzerland, 1981.
18. *Report on the Safety and Wholesomeness of Irradiated Foods*; Advisory Committee on Irradiated and Novel Foods. Department of Health and Social Security: London, 1986.
19. *Irradiation in the Production, Processing, and Handling of Food*; Final Rule, 51 Fed. Regist. 13376–13399; U.S. Government Printing Office: Washington, DC, April 18, 1986.
20. *Irradiation in the Production, Processing, and Handling of Food*; Final Rule, 55 Fed. Regist. 18538–18544; U.S. Government Printing Office: Washington, DC, May 2, 1990.
21. *Irradiation in the Production, Processing, and Handling of Food*; Proposed Rule, 49 Fed. Regist. 5714–5722; U.S. Government Printing Office: Washington, DC, Feb. 14, 1984.
22. *Document on Food Irradiation*: Adopted on 16 December 1988 by the FAO/IAEA/WHO/ITC-UNCTAD/GATT International Conference on the Acceptance, Control of, and Trade in Irradiated Food, World Health Organization: Geneva, Switzerland, 1988.

RECEIVED for review December 1, 1992. ACCEPTED revised manuscript March 25, 1993.

Cancer Incidence and Mortality after Iodine-131 Therapy for Hyperthyroidism

Per Hall and Lars-Erik Holm

Department of General Oncology, Radiumhemmet, Karolinska Hospital, Stockholm, Sweden

Cancer risk was studied in 10,552 Swedish hyperthyroid patients treated with ^{131}I between 1950 and 1975. Patients were followed for an average of 15 years (range 1–35 years) and were matched with the Swedish Cancer Register (SCR) and the Swedish Cause of Death Register (SCDR). The overall standardized incidence ratio (SIR) was 1.06 [95% confidence interval (CI) = 1.01–1.11], and the overall standardized mortality ratio (SMR) was 1.09 (95% CI = 1.03–1.16). The stomach was the only site for which cancer risk increased over time (p < 0.05) and with increasing activity of ^{131}I administered (p = not significant). No increased incidence of leukemia was found, which adds further support to the view that a radiation dose delivered gradually over time is less carcinogenic than the same total dose received over a short time. A possible excess owing to radiation was suggested only for stomach cancer.

IODINE-131 THERAPY FOR HYPERTHYROIDISM WAS FIRST INTRODUCED in the 1940s (*1, 2*) and, in many clinics, is considered to be the treatment of choice for this disease, largely because serious side effects are uncommon. However, concern still exists as to the possible carcinogenic effect of [131]I.

Reports on increased risks of breast (*3, 4*) and thyroid cancer (*3*) among hyperthyroid patients treated with ^{131}I are in contrast to others

0065–2393/95/0243–0103$08.00/0

that failed to detect increased cancer risks among such patients (5–8). In a recent study of 1762 hyperthyroid women, 80% of whom received ^{131}I, overall mortality was significantly elevated, but the standardized mortality ratio (SMR) for cancer did not differ from unity (4).

Although several studies of patients treated with ^{131}I were conducted, no clear pattern of risk was observed. Leukemia was never found to be in excess following ^{131}I therapy for hyperthyroidism. However, the risk of leukemia was elevated in three studies of thyroid cancer patients treated with larger doses of ^{131}I (9–11), but the number of leukemias was small in these studies with 2, 3, and 4 cases, respectively.

The purpose of the present study was to analyze the incidence and mortality of cancer and leukemia in a large Swedish population treated with ^{131}I for hyperthyroidism.

Subjects and Methods

The patients were admitted to seven university hospitals between 1950 and 1975 and were all under the age of 75 years at the time of first ^{131}I treatment. Ninety-four cases were excluded due to insufficient information on names and dates of birth. Mean age at the time of first ^{131}I treatment was 57 years (range 13–74 years).

Case records from the hospitals were used to obtain information on thyroid disorder and treatment. Some patients had previously received external radiotherapy toward the head–neck region (3%), thyroid hormone supplement for nontoxic goiter (2%), surgery for nontoxic goiter (3%), antithyroid drugs (24%), or surgery for hyperthyroidism (14%).

Fifty-nine percent of the patients received only one ^{131}I treatment, and 41% received two or more treatments. The mean total activity administered was 506 MBq (range 37–19,980 MBq). A total ^{131}I activity of 220 MBq or less (mean 150 MBq) was given to 30% of the patients, 221–480 MBq (mean 315 MBq) to 38%, and >480 MBq (mean 1063 MBq) to 32%. The mean number of treatments in each group was 1.1, 1.5, and 2.3, respectively.

The dose to the thyroid gland aimed at 60–100 Gy. In calculating mean organ doses, the International Commission on Radiological Protection (ICRP) tables (12), mean administered activity of ^{131}I, and mean 24-hour uptake of ^{131}I were used. The stomach wall received a mean dose of 0.25 Gy, and the urinary bladder wall and small intestine each received 0.14 Gy. No other organ received more than 0.10 Gy. The mean total body dose was estimated to be 0.08 Gy and the mean bone marrow dose to be 0.05 Gy.

The total cohort was matched with the Swedish Cancer Register (SCR) from 1958 to 1985 and the Swedish Cause of Death Register (SCDR) from 1952 to 1986 to identify cancer incidence and mortality in the cohort. The SCR receives notifications on newly diagnosed cancers, not only from clinicians but also from pathologists and cytologists. More than 96% of all cancers in Sweden are reported to the SCR (*13*). All deaths are certified by a physician. The matching concept for both record linkages was the unique identification number that is given to all individuals in Sweden.

Patients were considered to be at risk 1 year after the initial ^{131}I treatment until death or end of follow-up period. Attained age, sex, and calendar year were taken into consideration when the expected incidence and mortality were calculated using data from the SCR and the SCDR. The expected incidence and mortality were thus based on findings from the whole Swedish population (i.e., indirect standardization). Standardized incidence ratio (SIR) and SMR were defined as the ratio between observed and expected numbers and were calculated using the methods suggested by Breslow and Day (*14*). The 95% CI was determined by assuming the observed number of cases to be distributed as a Poisson variable.

Results

Within the first year of follow-up, 345 patients died, and the analyses were thus based on 10,207 patients. The patients were followed for an average of 15 years (range: 1–35 years).

A total of 1543 cancers were observed more than 1 year after exposure (SIR = 1.06; 95% CI = 1.01–1.11). More than 10 years after exposure, 830 cancers were seen (SIR = 1.10; 95% CI = 1.02–1.17; Table I), and significantly elevated risks were seen for cancer of the stomach (SIR = 1.33; 95% CI = 1.01–1.71; n = 58), kidney (SIR = 1.51; 95% CI = 1.06–2.08; n = 37), and nervous system (SIR = 1.63; 95% CI = 1.10–2.32; n = 30).

Deaths due to cancer or leukemia were observed in 977 cases. Fifty-three percent of the diagnoses were confirmed at autopsy, and an additional 46% were reported from hospitals but not based on autopsy findings. The overall SMR for malignant tumors and leukemia was 1.09 (95% CI = 1.03–1.16). Sites significantly elevated after 10 years of follow-up were cancer of the stomach (SMR = 1.41; 95% CI = 1.06–1.85; n = 54; Table I) and lung (SMR = 1.80; 95% CI = 1.39–2.31; n = 63).

A total of 37 leukemias occurred more than 1 year after exposure, and the SIR was 0.81 (95% CI = 0.57–1.12; Table II). Chronic lymphatic leukemia (CLL), a condition not known to be increased after

Table I. Observed Number of Cancers and Deaths from Cancer, 10
Years after Exposure, in 10,207 Hyperthyroid Patients Receiving ^{131}I, SIR,
SMR, and 95% CI for Selected Organs

Cancer Site	Incidence			Mortality		
	Obsd	SIR	95% CI	Obsd	SMR	95% CI
Stomach	58	1.33	1.01–1.71	54	1.41	1.06–1.85
Lung	50	1.17	0.87–1.54	63	1.80	1.39–2.31
Breast	134	1.04	0.87–2.33	39	0.77	0.54–1.05
Kidney	37	1.51	1.06–2.08	15	0.90	0.51–1.49
Bladder	28	1.13	0.75–1.63	8	0.71	0.31–1.40
Nervous system	30	1.63	1.10–2.32	8	0.97	0.42–1.91
Thyroid	9	1.32	0.61–2.50	2	0.66	0.08–2.37
Total[a]	830	1.10	1.02–1.17	510	1.14	1.04–1.24

[a]Also includes all sites not listed in the table.

Table II. Observed Number of Leukemias 1 Year
after Exposure in 10,207 Hyperthyroid Patients
Receiving ^{131}I, SIR, and 95% CI in Relation
to Type 1 of Leukemia

Type of Leukemia	Observed	SIR	95% CI
Non-CLL	25	0.85	0.55–1.25
CLL	12	0.75	0.39–1.30
All leukemias	37	0.81	0.57–1.12

irradiation, had approximately the same risk (SIR = 0.75) as non-CLL
(SIR = 0.85). Risk of leukemia did not vary by sex, age, time, or dose
of [131]I.

Table III shows the cancer incidence for some selected organs in
relation to follow-up. Except for stomach cancer there were no significant time trends for any of the cancer sites or for all cancers combined.

The mortality for stomach cancer and for all cancers combined increased with the increasing administration of ^{131}I activity (Table IV).
These trends, however, were not statistically significant. No trend was
seen for lung cancer, although the highest risk was seen in patients
receiving ≥481 MBq. Similar patterns were seen when incidences were
analyzed. The elevated risk for thyroid cancer (n = 8) among patients
given ≥481 MBq included six thyroid cancers diagnosed during the
first 5 years of follow-up.

Discussion

Patients receiving ^{131}I therapy for hyperthyroidism had an overall cancer incidence (6%) and cancer mortality (9%) slightly greater than ex-

Table III. Observed Number of Deaths from Cancer 1 Year after Exposure in 10,207 Hyperthyroid Patients Receiving ^{131}I, SIR, and 95% CI in Relation to Duration of Follow-up

| | Years of Follow-up | | | | | | | | |
| | 1–9 | | | 10–14 | | | >15 | | |
Cancer Site	Observed	SIR	95% CI	Observed	SIR	95% CI	Observed	SIR	95% CI
Stomach	34	0.77	0.53–1.08	23	1.10	0.70–1.65	35	1.54	1.07–2.14
Lung	55	1.49	1.12–1.93	20	0.99	0.61–1.53	30	1.33	0.90–1.90
Breast	135	1.01	0.85–1.20	64	1.01	0.78–1.29	70	1.06	0.82–1.33
Kidney	29	1.26	0.85–1.81	23	1.95	1.24–2.93	14	1.10	0.60–1.83
Bladder	23	1.11	0.70–1.66	14	1.22	0.67–2.05	14	1.05	0.58–1.76
Nervous system	18	0.89	0.53–1.41	17	1.74	1.02–2.78	13	1.30	0.69–2.21
Thyroid gland	9	1.25	0.57–2.38	5	1.49	0.49–3.48	4	1.15	0.32–2.39
Total^a	713	1.02	0.95–1.10	394	1.10	0.99–1.21	436	1.10	1.00–1.20

^{a}Includes cancer sites not listed in the table.

Table IV. Observed Number of Deaths from Cancer 1 Year after Exposure in 10,207 Hyperthyroid Patients Receiving ^{131}I, SMR, and 95% CI in Relation to Administered Activity

Cancer Site	≤220 MBq			221–480 MBq			≥481 MBq		
	Observed	SMR	95% CI	Observed	SMR	95% CI	Observed	SMR	95% CI
Stomach	25	1.00	0.65–1.48	35	1.05	0.73–1.46	33	1.22	0.84–1.72
Lung	33	1.20	0.83–1.69	33	0.95	0.66–1.34	41	1.56	1.12–2.11
Breast	20	0.58	0.35–0.89	46	1.08	0.79–1.44	28	0.87	0.57–1.25
Female genital organs	36	1.19	0.83–1.65	41	1.11	0.79–1.50	37	1.31	0.92–1.81
Male genital organs	5	0.62	0.20–1.44	18	1.68	0.99–2.65	8	0.86	0.37–1.69
Kidney	12	1.17	0.61–2.05	20	1.53	0.94–2.37	13	1.30	0.69–2.22
Bladder	5	0.82	0.26–1.90	9	1.12	0.51–2.13	4	0.63	0.17–1.62
Nervous system	6	1.01	0.37–2.20	4	0.56	0.15–1.44	4	0.76	0.21–1.96
Thyroid	2	1.08	0.13–3.91	3	1.25	0.26–3.64	8	4.21	1.82–8.30
Lymphomas	5	0.66	0.21–1.54	10	1.04	0.50–1.91	4	0.54	0.15–1.40
Total[a]	274	0.98	0.86–1.10	388	1.09	0.98–1.19	315	1.13	1.01–1.26

[a]Also includes all sites not listed in the table.

pected in the general population. Only stomach cancer revealed both increased incidence and mortality figures and was also the only site where the risk increased with increasing dose. The stomach wall received the highest dose, 0.25 Gy, and it is possible that the ^{131}I explains the excess.

The increased medical surveillance of the patients could have led to detection of cancers not causing symptoms, thus indicating a higher risk for cancer. Many studies showing elevated cancer risks after exposure to low doses of ionizing radiation have been based on studies of children. Only 5% of the patients in the present study were younger than 40 years at the time of ^{131}I exposure. The risk associated with protracted whole-body exposure of 0.08 Gy from ^{131}I may not be comparable to the high dose rate situations in other studies in which increased cancer mortality was observed. The latency period for solid tumors is usually at least 10 years (*15, 16*). The mean follow-up period was 18 years among the 7818 patients surviving 10 years, and it is possible that this follow-up period was not sufficiently long to detect an increased cancer mortality due to radiation exposure.

The study population represented a select group, as young patients and women of fertile age generally were not likely to receive ^{131}I because of the potential hazards of such therapy. The study group also consisted of patients unfit for surgery because of cardiac or respiratory diseases. Some of the risk factors for these diseases, such as diet and smoking, might have influenced the cancer risk. A reference population of nonirradiated patients with hyperthyroidism would have been preferable, instead of the country as a whole, even if this method also would have included a selection bias because there was always a reason why some patients were given radiotherapy and others not. The strengths of the study were the few patients lost to follow-up, the accuracy of the SCR and SCDR, and data on individually administered ^{131}I activity.

The puzzling finding that more patients died of lung cancer ($n = 63$) than received the diagnosis ($n = 50$) is explained by the fact that the autopsy often defined a more specified diagnosis than was indicated in the clinical records. It was shown that when an autopsy, disproving the original diagnosis, was delayed, the death certificate was not amended (*17*).

The SMR for cancer of the stomach and lung was highest among those receiving the highest ^{131}I activity. In a study by Darby et al. (*18*) patients receiving radiotherapy for ankylosing spondylitis, a benign condition of the spine, showed increased mortality from esophagus, stomach, and lung cancer. These organs received >10 Gy, and no decreasing risk with increasing age at exposure was noticed. A selection bias could partly explain our findings because the prevalence of smokers is expected to be higher among patients with hyperthyroidism than in the normal population (19).

In a study of thyroid cancer patients (11), an elevated risk for stomach cancer was also found in those treated with ^{131}I in contrast to thyroid cancer patients receiving other types of treatment. The risk increased over time, and the mean dose to the stomach was 2.1 Gy compared with 0.25 Gy in the present study.

A mean dose of approximately 0.5 Gy to the thyroid gland was received by 35,000 patients examined with [131]I (20). No increased risk related to the exposure was found. The dose to the thyroid in the present study aimed at 60–100 Gy, and it is likely that this dose had a cell-killing instead of a carcinogenic effect because no increased mortality was seen in patients followed for more than 10 years after exposure. Because the cure rate of thyroid cancer is high, incidence data instead of mortality data should be used. However, no significantly increased incidence of thyroid cancer was noticed.

Patients receiving the highest ^{131}I activity had the highest cancer mortality and probably also the most severe hyperthyroidism, as reflected by the higher number of treatments. In a recent study of the present cohort, elevated risks for most causes of death were found among those receiving the highest amount of [131]I (21). It was concluded that the underlying disease rather than the ^{131}I therapy was the reason for this. If these patients also had a cancer, although not the cause of death, they would have been reported as dying from a malignant disease. This probably contributes to our findings of a slightly increased overall risk in the group receiving the highest ^{131}I activity.

The induction of leukemia by ionizing radiation has been well documented, and excess mortality seems to reach a peak within 10 years after exposure (15, 16). Among the atomic bomb survivors, elevated risks were found among those with an absorbed dose to the bone marrow of >0.5 Gy but not among those with absorbed doses lower than 0.5 (22). In patients treated with X-ray for ankylosing spondylitis, elevated risks of leukemia were found (18), and excess mortality became detectable within 2 years of exposure and peaked within the first 5 years. In our study 37 leukemias were found more than 1 year after exposure, and SIR was 0.81. Using data from the atomic bomb survivors (22) of an excess relative mortality per Gy organ-absorbed dose of 5.21 and an estimated bone marrow dose of 0.05 Gy, the SMR in our study would be 1.26. The lack of correspondence is probably explained by the large difference in dose rate, because the biological half-life of ^{131}I is at least 8 days.

Our observations suggest that (a) low doses of ionizing radiation are less effective in inducing cancer than higher doses, (b) the protracted nature of the ^{131}I exposure makes the isotope a less effective carcinogen probably because of the opportunity of cellular repair of the radiation damage, (c) the dose was so low that a detectable increase in cancer or leukemia was unlikely, and (d) extrapolating from high doses to low doses and dose rates does not seem to underestimate the risk.

Acknowledgments

This study was performed in cooperation with the following: G. Lundell, Department of General Oncology, and K. Wiklund, Department of Cancer Epidemiology, Radiumhemmet, Karolinska Hospital, Stockholm, Sweden; M. Lidberg, Department of Hospital Physics, South Hospital, Stockholm, Sweden; E. Cederquist and J. Tennvall, Department of General Oncology, University Hospital, Lund, Sweden; G. Bjelkengren, Department of General Oncology, and U.-B. Ericsson, Department of Internal Medicine, Malmö General Hospital, Malmö, Sweden; G. Berg, Department of General Oncology, and S. Lindberg, Nuclear Medicine Division, Sahlgren's Hospital, Gothenburg, Sweden; H. Wicklund, Department of General Oncology, University Hospital, Uppsala, Sweden; A. Hallquist and L.-G. Larsson, Department of General Oncology, University Hospital, Umeå, Sweden; and J. D. Boice, Jr., Epidemiology and Biostatistics Program, Division of Cancer Etiology, National Cancer Institute, Bethesda, MD.

The study was supported by Public Health Service Contract No. N01-CP-51034 from the National Cancer Institute, Bethesda, MD.

The authors wish to thank Elisabeth Bjurstedt for valuable assistance in various aspects of our study.

References

1. Hamilton, J. G.; Lawrence, J. H. *J. Clin. Invest.* **1942**, *21*, 624.
2. Hertz, S.; Roberts, A. *J. Clin. Invest.* **1942**, *21*, 624.
3. Ron, E.; Curtis, R.; Hoffman, D. A.; Flannary, J. T. *Br. J. Cancer* **1984**, *49*, 87–92.
4. Goldman, M. B.; Maloof, F.; Monson, R. R.; Aschengrau, A.; Cooper, D. S.; Ridgway, E. C. *Am. J. Epidemiol.* **1988**, *127*, 969–980.
5. Hoffman, D. A.; McConahey, W. M. *J. Natl. Cancer Inst.* **1983**, *70*, 63–67.
6. Hoffman, D. A. In *Radiation Carcinogenesis: Epidemiology and Biological Significance*; Boice, J. D., Jr.; Fraumeni, J. F., Jr., Eds.; Raven Press: New York, 1984; pp 273–280.
7. Holm, L.-E. In *Radiation Carcinogenesis: Epidemiology and Biological Significance*; Boice, J. D., Jr.; Fraumeni, J. F., Jr., Eds.; Raven Press: New York, 1984; pp 263–271.
8. Holm, L.-E.; Hall, P.; Wiklund, K.; Lundell, G.; Berg, G.; Bjelkengren, G.; Cederquist, E.; Ericsson, U.-B.; Hallquist, A.; Larsson, L.-G.; Lid-

berg, M.; Lindberg, S.; Tennvall, J.; Wicklund, H.; Boice, J. D., Jr. *J. Natl. Cancer Inst.* **1991**, *83*, 1072–1077.

9. Brincker, H.; Hansen, H. S.; Andersen, A. P. *Br. J. Cancer* **1973**, *28*, 232–237.
10. Edmonds, C. J.; Smith, T. *Br. J. Radiol.* **1986**, *59*, 45–51.
11. Hall, P.; Holm, L.-E.; Lundell, G.; Bjelkengren, G.; Larsson, L.-G.; Lindberg, S.; Tenvall, J.; Wicklund, H.; Boice, J. D., Jr. *Br. J. Cancer* **1991**, *64*, 159–163.
12. International Commission on Radiological Protection; *Radiation Dose to Patients from Radiopharmaceuticals*; Annals of the ICRP: Publication 53; Pergamon Press: Oxford, United Kingdom, 1988; Vol. 18.
13. Mattsson, B.; Wallgren, A. *Acta Radiol. Oncol.* **1984**, *23*, 305–313.
14. Breslow, N. E.; Day, N. E. In *The Design and Analysis of Cohort Studies*; International Agency for Research on Cancer: Lyon, France, 1987; Vol. II.
15. United Nations Scientific Committee on the Effects of Atomic Radiation; *Sources, Effects and Risks of Ionizing Radiation*; 1988 Report to the General Assembly, with annexes; United Nations: New York, 1988.
16. Committee on the Biological Effects of Ionizing Radiations; *Health Effects of Exposure to Low Levels of Ionizing Radiation*; BEIR V; National Academy Press: Washington, DC, 1990.
17. Mattsson, B.; Rutqvist, L.-E. *Radiother. Oncol.* **1985**, *4*, 63–70.
18. Darby, S. C.; Doll, R.; Gill, S. K.; Smith, P. G. *Br. J. Cancer* **1987**, *55*, 179–190.
19. Bartalena, L.; Martino, E.; Marcocci, C., et al. *J. Endocrinol. Invest.* **1989**, *12*, 733–737.
20. Holm, L.-E.; Wiklund, K. E.; Lundell, G. E., et al. *J. Natl. Cancer Inst.* **1989**, *81*, 302–306.
21. Hall, P.; Lundell, G.; Holm, L.-E. *Acta Endocrinol.* **1993**, *128*, 230.
22. Shimizu, Y.; Kato, H.; Schull, W. J. *Radiat. Res.* **1990**, *121*, 120–141.

RECEIVED for review October 2, 1992. ACCEPTED revised manuscript March 25, 1993.

Effects of Large Doses of Radiation

The Genetic Effects of Human Exposures to Ionizing Radiation

James V. Neel

Department of Human Genetics, M4708 Medical Science II 0618, University of Michigan, Ann Arbor, MI 48109–0618

The data on the potential genetic effects of atomic bombs collected over the course of the past 46 years are reviewed. No statistically significant effect of parental exposure to the ionizing radiation of the bombs on the frequency of congenital malformations, stillbirths, survival, physical growth and development, malignant tumors with onset prior to age 20, certain chromosome abnormalities, or mutations involving the structure and function of a battery of proteins was found. The effect of exposure averaged over all indicators, however, is slightly positive. This finding appears to be the current best estimate of the genetic effect of the exposure of humans to ionizing radiation. From the data on the control children in this study, the contribution of spontaneous mutation each generation to each of these indicators was estimated, and we calculated the amount of acute ionizing radiation necessary to increase this spontaneous rate by 100% (the so-called doubling dose). This is estimated to be between 1.7 and 2.2 Sv equivalents, with an error term difficult to estimate. It is argued that, for exposures to chronic ionizing radiation, the doubling dose is approximately 3.4–4.4 Sv equivalents, again with an error term extending well beyond this interval. Our review of the experimental data from mice with respect to eight specific locus and phenotype systems suggests good agreement with the estimate for humans. This estimate is about four times higher than the estimate employed in most past treatments of the genetic risks of exposure to ionizing radiation.

0065–2393/95/0243–0115$08.00/0

Since 1946 a continuous effort has been underway to collect data on the potential genetic effects of the atomic bombs detonated over Hiroshima and Nagasaki in 1945, as well as data on other potential delayed effects of the bombings. This has involved a joint U.S.–Japanese effort, the principal organization involved on the U.S. side being the National Academy of Sciences, with funds from the Department of Energy, and on the Japanese side, their Ministry of Welfare. The study has been the labor of many people. In this chapter I propose the following: first, to describe the results of these studies; second, to present our effort to estimate a "genetic doubling dose" from these data; third, to compare these findings and estimates with the extensive body of data on the genetic effects of radiation available for mice; and, fourth, to contrast these results with the results of the recent, so-called Sellafield study (1). Because the intellectual background of this readership is primarily chemical rather than biological, I will avoid abstruse genetic terminology where possible but, at some points, will necessarily be rather technical.

Design of Study

Over the years a roster was established of all the children born in Hiroshima and Nagasaki to parents one or both of whom were within the zone of significant radiation at the time of the bombings (ATB), the so-called proximally exposed parents. Two matching rosters with respect to city, sex, and year of birth were also established. One roster was composed of children at least one of whose parents was in Hiroshima or Nagasaki ATB but sufficiently far from ground zero that they received no increased radiation ATB, and the other roster was composed of children whose parents were not in either city ATB but moved in subsequently (2–4). The zone of significantly increased radiation exposure from the bombs did not extend beyond 2500 m from the hypocenter; the position of the parents with respect to the hypocenter and their shielding ATB were the primary bases for the estimation of the precise radiation exposure of each parent. Now, 47 years after the bombings, the roster of children ever to be born to proximally exposed survivors is essentially complete, totaling approximately 31,000 children, and all the rosters are closed. I shall refer to each of these rosters as a cohort. Because these children were registered at birth—and, indeed, in the early years of the program (1948–1954), registered prebirth, when for ration purposes the mothers registered their pregnancies—this was a *prospective* study, an important methodological point.

These three cohorts were studied in various ways. At the outset attention centered on congenital malformations, stillbirths, and neonatal deaths, which separately or in combination are referred to as an "untoward pregnancy outcome" (UPO). Sex of the child, of course, was recorded. The program that evaluated congenital malformations at birth and at age 9 months was abandoned in 1954, but all of these children plus subsequent accessions to the cohorts were followed with respect to survival of the live-born children and with respect to the diagnosis of a malignant tumor. A subset of the live-born children in the cohorts who survived to age 13 was examined for the occurrence of cytogenetic abnormalities, such as an abnormal number of sex chromosomes or an exchange of segments between (or within) chromosomes. Another subset, overlapping in part with the previous subset and also restricted to children reaching age 13, was studied for the occurrence of mutations altering the electrophoretic behavior or activity of a battery of 30 proteins found in serum plasma or red blood cells. Finally, there are data on birth weights for the children examined for an untoward pregnancy outcome, data on physical development at 9 months on a subset of these children, and data on development at ages 6–17 for a further subset that overlaps with the first two. I emphasize that, from the outset of these studies, a major effort was made to respect Japanese sensibilities.

Summary of Findings

The data that result from these examinations are basically of three types. With respect to UPO, death of live-born infants, and cancer before the age of 20, there are many causes, of which mutation in the preceding generation is only one. Furthermore, with respect to the evaluation of mutational damage, we cannot determine which among the individuals who exhibit a congenital defect or fail to survive or develop a malignancy owes this to the atomic bomb-induced mutations whose frequency we are seeking to measure and which ones owe the finding to some other cause. We must for these data obtain the regression of the indicator on radiation exposure and then attempt to estimate the fraction of the control data that results from spontaneous mutation in the preceding generation, that being the fraction expected to increase in consequence of the radiation exposure and to which the regression term applies. For the cytogenetic abnormalities and protein variants, on the other hand, we can by the appropriate family studies determine exactly which individuals exhibiting the trait are mutants, whose occurrence should be analyzed in relation to parental exposure to radiation. Finally, studying physical development is essentially a matter of comparing the means and variances of the several measurements in relation to parental exposure history, on the thesis that an ıutation rate in the exposed parents will be reflected in physical development of their children.

presenting the data, it is important to convey some idea iation exposures involved. In the early 1980s the question of the ᴗ.gan doses sustained by survivors underwent a thorough re-evaluation, resulting in what is called the Dose Schedule 1986 (DS86). All of the genetic data were analyzed on the basis of radiation exposures to the testes or ovaries, as estimated by the DS86 schedule. The radiation from the bombs was predominantly gamma in type, akin to the exposure from X-rays, but there was a small neutron component. Neutron radiation is more effective than gamma radiation in producing genetic effects; on the basis of the experimental literature (5), we assigned the neutron component a relative biological efficiency (RBE) of 20 for genetic effects.

Because the spectrum of radiation is mixed neutron and gamma, doses to the testes and ovaries must be expressed in sieverts (Sv) [1 Sv equivalent equals 100 roentgen equivalent units (rem)]. In general, the average conjoint parental gonad exposure for those parents receiving increased radiation ATB is about 0.4–0.5 Sv equivalents. This is not a large dose by the standards of the experimental geneticist who works with radiation. We appreciated from the beginning of our study the fact that gonadal radiation doses would be much smaller than those employed by experimentalists and realized that we would be searching for small effects. This fact made it very important that the study be structured so that the parents who did not receive increased radiation at the time of the bombings be as well matched as possible to the parents who did receive increased radiation.

We now proceed with the discussion of our data. The regression on conjoint parental dose (Sv equivalents) of UPOs was 0.00264 ± 0.00277 and of mortality (exclusive of cancer) of live-born infants through an average age of 26 years, 0.00076 ± 0.00154 (6, 7). Each of these regressions is based upon all the data of the three cohorts. The regression term is derived from a simple linear dose–response model, where the background characteristics to be factored in usually include city, sex of child, and age of the father and mother, and the risk factors are, usually, conjoint parental gonad dose, birth year, and (where available) consanguinity of parents. Both regressions are positive but smaller than their standard errors (i.e., do not approach statistical significance).

Because of a recent report that will be discussed later, particular attention will be paid to the data on cancer incidence (8). As shown in Table I, there were 92 reports of malignant tumors diagnosed prior to age 20 in 72,216 subjects for whom the radiation exposures of their parents could be estimated. For purposes of analysis, these reports

Table I. Results of a Linear Multiple, Least-Squares Regression Analysis of the Incidence of Cancer below the Age of 20, by Conjoint Parental Dose (Sv, RBE = 20), City, Sex, and Birth Years (72,216 Subjects)

Category	Conjoint Dose (Sv, RBE = 20)	Hiroshima	Maleness	Years since Birth	Intercept[a]
All cancers (92 cases)	-0.000081 ±0.000252	-0.000247 ±0.000275	0.000602[b] ±0.000266	0.000004 ±0.000017	0.001134 ±0.000260
Heritable (19 cases)	-0.000073 ±0.000114	-0.000156 ±0.000125	-0.000040 ±0.000121	0.000007 ±0.000008	0.000395 ±0.000118
Leukemia (33 cases)	0.0000003 ±0.000151	0.000075 ±0.000165	0.000281 ±0.000159	-0.000007 ±0.000010	0.000266 ±0.000156
Other (40 cases)	-0.000009 ±0.000166	-0.000165 ±0.000181	0.000361[b] ±0.000175	0.000004 ±0.000011	0.000473 ±0.000171

[a] Adjusted for the average years between bombing and birth.
[b] $0.01 < p < 0.05$.

were subdivided into three categories. First are the 19 cancers for which there is currently evidence of a familial genetic basis, that is, cancers such as retinoblastoma, Wilms tumor, and neuroblastoma, which collectively can be termed "heritable". For these cancers a predisposing factor is often a transmitted mutation from one of the parents. The second category is the leukemias, which in humans have not shown the inherited genetic predisposition of the first group of tumors (33 cases). The final category is composed of the 40 remaining cancers. The regressions for these three categories and for the combined data are shown in Table I. No evidence for any of these categories of a radiation effect is shown; the regressions all lay well within the error terms. For purposes of comparison, we include in the table the effects of some of the other variables included in the analysis. The sensitivity of the analysis is indicated by the fact that the well-established, slightly greater prevalence of cancer in young males than in young females emerges in this study at the 5% probability level.

Table II summarizes the data on abnormalities in the number of sex chromosomes, termed "sex-chromosome aneuploidy" (9). The frequency per 1000 children examined was 2.28 for the children of radiation-exposed parents and 3.01 for those of nonexposed parents. All sex-chromosome aneuploids are presumed to be caused by a chromosomal mutation in a parent. This slightly lower frequency in the children of exposed parents is counter to the direction of genetic expectation but, of course, is not statistically significant.

Table III presents the data on rearrangements of the chromosomes of the type termed "balanced", in which there is no net gain or loss of genetic material (9). In this case, family studies must be performed to determine whether the abnormality was inherited or is due to a mutation, a de novo event. As Table III shows, there was just one de novo abnormality in the children of exposed and one in the children of unexposed controls. Although we did not feel such a small number of positive findings warranted a regression analysis, clearly there is no hint of a significant radiation effect.

Table II. Sex Chromosome Abnormalities in the Children of Radiation-Exposed and Unexposed Survivors of Hiroshima and Nagasaki[a]

City	Group	Number Studied	Abnormalities	Rate per 1000
H	Exposed	4716	12	2.54
	Control	5112	17	3.33
N	Exposed	3606	7	1.94
	Control	2864	7	2.44
H + N	Exposed	8322	19	2.28
	Control	7976	24	3.01

[a]All children were examined subsequent to their 13th birthday.

Table III. Frequency and Parent of Origin of Balanced Structural Rearrangements[a] in the Cytogenetic Study of Awa et al. (9)

	Hiroshima		Nagasaki		H + N	
	Exposed	Control	Exposed	Control	Exposed	Control
Studied for parent of origin						
De Novo (mutation)	1	1	0	0	1	1
Inherited						
Father	2	4	2	4	4	8
Mother	0	1	0	1	0	2
Undetermined	6	3	0	2	6	5
Subtotal	9	9	2	7	11	16
Not studied for parent of origin	5	7	2	2	7	9
Grand total	14 .	16	4	9	18	25
Children studied	4716	5112	3606	2864	8322	7976

[a]These data include reciprocal and Robertsonian translocations plus pericentric inversions.

Table IV. A Summary of the Search for Mutations Altering the Electrophoretic Mobility of Proteins or the Activity of Enzymes[a]

	Proximally Exposed Parents	Distally Exposed Parents	Total
Electromorphs			
Number of children examined	13,052	10,609	23,661
Equivalent locus tests	663,494	466,881	
Mutations	3	3	6
Mutation rate per locus per generation	0.45×10^{-5}	0.64×10^{-5}	0.53×10^{-5}
Enzyme deficiency variants			
Number of children examined	4,989	5,026	10,015
Equivalent locus tests	60,529	61,741	122,270
Mutations	1	0	1
Mutation rate per locus per generation	1.65×10^{-5}	0	0.82×10^{-5}

[a]Further explanation exists in text.

Table IV summarizes the data on mutations affecting the electrophoretic mobility or the activity of a series of 30 proteins (10). The controls are again drawn from the children of the two groups of unirradiated parents. With respect to the electrophoretic mutations, there are three in the children of exposed parents and three in the children of parents receiving no excess radiation. With respect to enzyme deficiency variants, the score is one in the children of radiation-exposed parents and zero in the children of radiation-unexposed parents. When these findings are combined, the mutation rates are almost identical for the two groups (0.55×10^{-5} for the children of radiated and 0.59×10^{-5} for the children of nonradiated parents). When the mutations involving proteins and sex-chromosome aneuploids are distributed according to parental radiation exposures and a regression is obtained, it is 0.00001 ± 0.00001/Sv equivalent for the proteins and 0.00044 ± 0.00069/Sv equivalent for the aneuploids.

With respect to a seventh indicator to which attention was directed in the past, the sex ratio, we now feel that, because of technical considerations that will not be discussed in this chapter, the only useful data are those on the sex ratio in the children of radiated mothers. These data (not shown) reveal a nonsignificant increase in male births in the children of exposed mothers (11); that is, these data are not in the direction of expectation if radiation caused an increase in mutations occurring on the sex chromosomes of these exposed mothers. Likewise (data not shown), based on measurements at birth, at 9 months, and at school age, there is no evidence of impaired physical development in the children of exposed parents (2, 12–16).

Inferences To Be Drawn from These Data

In short, we cannot demonstrate in the strict statistical sense that parental exposure to the bombs adversely influenced those attributes of their children that we were able to measure. The differences observed all could have occurred by chance. Neither, on the other hand, because of the statistical uncertainties surrounding a study of this nature, can we exclude a small effect but one of which society needs to be aware. Because of the relatively low radiation exposures to the germ cells—small by the standards of the experimentalist—we were aware of the possibility of an ambiguous outcome from the outset of the studies and have devoted a great deal of thought to how best to use these data for the public welfare. In this kind of situation, the data can be considered at two levels. On one hand, the various studies may be regarded as empirical exercises, a body of data to be referred to whenever the question of the genetic impact of radiation on a population arises, with no further attempt at interpretation. In this context we want to emphasize that the end points we have studied are those of major human concern: congenital defects, premature death, chromosomal abnormalities associated with sterility or setting the stage for defective children in the next generation, and cancer.

However, an alternative approach to this situation exists. We do not regard these studies as testing the hypothesis that radiation produces mutations. This hypothesis has held true in every plant and animal species that has been properly studied with regard to this question. Therefore, we can treat the observed results as our best current estimate of the genetic impact of radiation on humans and consider how to analyze these results to extract the best guidelines possible concerning the genetic effects of human exposures to ionizing radiation. In this regard, we point out that, for two of the five indicators for which regressions were derived, the regression on radiation exposure is actually negative. For these indicators, the differences between the children of exposed parents and controls are in fact very close to zero; the occurrence of one or two positive outcomes in the children of the more heavily exposed parents would create an excess frequency of positive outcomes in the children of the exposed parents, that is, a positive regression. What this suggests is that, in this population, the genetic effects of radiation are sufficiently small so that, owing to a combination of sample size, genetic contribution to indicator, and sensitivity of that indicator to radiation, by chance, these counter-hypothesis observations may occur.

Before we can use the data to generate an estimate of the doubling dose, we must estimate the contribution of each generation of mutation in the preceding generation to the indicators we have pursued. With respect to this estimate, the indicators fall into two categories. For two indicators, chromosomal abnormalities and the protein variants, the mutational contribution can be derived from the appropriate family studies. As regards the role of spontaneous mutation in the parental generation in UPOs, in early mortality, and in childhood cancer, a more indirect approach is necessary. As mentioned earlier, we cannot determine which specific affected children owe their condition to mutation. We can, however, drawing on the genetic literature, estimate the proportion of all such outcomes due to spontaneous mutation in the preceding generation. The technicalities of this effort are rather abstruse and will not be discussed here. As shown in Table V, elsewhere (17) we have suggested that in absolute terms between 0.0033 and 0.0053 of all pregnancy outcomes are characterized by an untoward event or prereproductive death presumed due to spontaneous mutation in the preceding generation. This is the fraction of pregnancy outcomes that should increase if radiation is responsible for an elevated mutation rate.

Since the observed total background frequency of these five outcomes in the children of the unirradiated parents was 0.102, this is comparable to the suggestion that 3–4% of these events can in the pregnancy terminations of the unexposed Japanese population be attributed to spontaneous mutation in the parental generation. This figure includes the contribution of the protein mutations to UPOs and to the survival of live-born infants, as recorded in this study. Sankaranarayanan (18; see also 19) has faulted the analysis for including the

Table V. A Summary of the Regression of the Various Indicators
on Parental Radiation Exposure and the Impact of Spontaneous Mutation
on the Indicator

Trait	Regression per Combined Parental (Sv)	Contribution of Spontaneous Mutation
UPO	+0.00264 ±0.00277	
F_1 mortality	+0.00076 ±0.00154	0.0033–0.0053
Protein mutations	−0.00001 ±0.00001	
Sex-chromosome aneuploids	+0.00044 ±0.00069	0.0030
F_1 cancer	−0.00008 ±0.00028	0.00002–0.00005
	0.00375	0.00632–0.00835

nongenetic causes of UPOs and early death in this baseline figure of
the impact of spontaneous mutation; he is in error. With respect to
the remaining mutational damage detected in this study, there is an
additional relatively large contribution by the sex chromosome aneuploids as well as a relatively small contribution of parental mutation
to childhood cancer. The total impact of newly arisen spontaneous mutation on these indicators is placed at 0.00632 to 0.00835.

Since all the regressions evaluate the effects of essentially non-
overlapping indicators (i.e., are independent of each other) and are
based on the same total material or subsets thereof, it is legitimate to
combine these additively; the sum of the individual regressions is
0.00375/Sv equivalent. The error term, of course, is relatively large
for each of the individual regressions, but collectively these regressions constitute a considerable body of data.

A convenient way to express the genetic effects of radiation is as
a doubling dose. This is the amount of radiation that will produce the
same frequency of mutation as occurs spontaneously each generation.
In these data the estimated total contribution from spontaneous mutation in the preceding generation to the sum of the various end points
we have pursued is between 0.0063 and 0.0084 (Table V). The combined regression term is +0.00375/Sv equivalent. The doubling dose
is simply the figure obtained by dividing the total mutational contribution to the indicators by the summed regressions per Sv equivalent.
With these input values, the doubling dose is estimated at between
1.68 and 2.22 Sv equivalents (17). From these various regressions we
have calculated that the minimal doubling dose at the 95% probability
level that results from pooling the estimates for UPOs, F_1 cancer, and
F_1 mortality is 0.63–1.04 Sv equivalents; the similar estimate for the

sex-chromosome aneuploids and protein mutations is 2.71 Sv equivalents (*17*).

With allowance for the errors in the individual regressions, these limits, of course, would be even wider. *We emphasize that this estimate is time and place specific and carries a considerable but indeterminate error.* In particular, the infant mortality was substantially higher in the immediate postwar years in Japan, when the bulk of the data were collected, than it is now, when it is very similar to the infant mortality in Europe and the United States. This implies that, in the early years of the study, selection against genetically determined disease resulting in early death was higher than at present. Caution in extrapolating from these results to other populations is indicated. On the other hand, inasmuch as the findings on several indicators that for technical reasons were not incorporated into the analysis (reciprocal chromosomal translocations, sex ratio, growth, and development) also failed to suggest a radiation effect, we feel our estimate is conservative.

Most of the radiation humans experience is not high dose rate, short duration exposures resulting in gonadal doses between 0.01 and 3.0 Sv equivalents, as in Hiroshima and Nagasaki, but low level, intermittent, or chronic exposure. The most extensive body of data on the genetic yield of low dose-rate exposures is derived from the well-known mouse 7-locus test system of Russell (*20*), where, at low dose rates, only one-third as many mutations are recovered as at high rates (*21*). Some of the other test systems in mice have yielded dose-rate factors of 5–10. Because the Japanese gonadal exposures were so much lower than the gonadal exposures employed in the experiments with mice, we believe it is appropriate to apply a dose-rate factor of 2 in extrapolating from the results of this study to the cumulative effects of intermittent or chronic low-level exposures (detailed argument in reference *17*). Then the doubling dose for the kind of radiation exposures we receive occupationally, medically, or naturally becomes approximately 4.0 Sv equivalents.

A Contrast of These Findings with the Results of Studies on Mice

For the past 40 years our thinking about the genetic implications of radiation has been guided largely by the results of experiments with mice, as it should have been while the human data were coming in. Furthermore, there are aspects of human radiation risks that for the foreseeable future will continue to be guided by experimental results. I refer especially to the demonstration that an increased mutation rate is not observed in the late litters born to irradiated female mice (22) and the demonstration, already mentioned, that a given dose of radiation administered in small accretions or chronically has roughly one-third to one-fifth or perhaps even less the genetic impact of the same dose administered acutely (21). However, now that the human data are in, the view of the genetic risks of radiation just presented differs to a nontrivial degree from that which was extrapolated to humans from the mouse data and incorporated into numerous national and international evaluations of radiation risks (cf. references 23 and 24). This extrapolated doubling dose for acute radiation has commonly been set at 0.4 Sv equivalents, and for chronic radiation, at 1.0 Sv equivalents. The most recent report of the U.S. National Research Council Committee on the Biological Effects of Ionizing Radiation (BEIR V), issued when the analysis of the Japanese data was still incomplete, moves somewhat in the direction of our findings, suggesting, for chronic radiation, a doubling dose *not less* than 1.0 Sv equivalents (25). Taken at face value, our estimate of the doubling dose for chronic radiation is approximately four times greater than that usually projected for humans from experiments on mice. This means the genetic risk of radiation, if we are correct, is four times less than the estimate previously projected from the mouse data. This constitutes a rather major revision in our understanding of the genetic effects of radiation.

This discrepancy led Lewis and me to attempt 2 years ago a point-by-point comparison of the data on mouse radiation genetics with the Hiroshima–Nagasaki data (26). We soon found such a comparison to be extremely difficult, because (1) mice are born in litters in which pre- and postnatal competition must be severe, (2) mice are born at a developmental age roughly equal to a 100-day-old human fetus, and (3) certain data available for humans are lacking in mice (and vice versa). The most appropriate comparison we could generate involved contrasting the results of all the specific locus–specific phenotype tests carried out thus far on mice with the estimate we generated for humans. The simple, unweighted average of the estimates of the doubling dose for acute ionizing radiation yielded by eight different specific locus–specific phenotypes systems was 1.35 Gy. These results are derived from experiments with male mice; the data do not exist to make an adjustment for a bisexual population. Following the precedent set by geneticists working with mice, it seems appropriate to apply a dose-rate factor of 3 to the mouse data when extrapolating from the relatively high acute exposures of mice to the doubling dose of chronic radiation. The result is an estimate of 4.05 Gy, in excellent agreement with the human data.

How could Lewis and I derive so different an estimate than that generated, for example, by the United Nations Scientific Committee on the Effects of Atomic Radiation (UNSCEAR) (*23; see also 27*)? First, the eight specific locus and phenotype data sets we employed included some data either not available to, or ignored by, the committee. Second, we felt one of the indicators employed in the UNSCEAR estimate, semisterility, was inappropriate for humans. But, finally, we did not place the heavy emphasis the committee did on the results of the well-known Russell 7-locus test. Elsewhere, we have argued that, in developing this system, Russell inadvertently selected for genes with relatively high mutability (*26*).

My intention in this chapter is not to overemphasize the apparent correspondence between the doubling-dose estimates for humans and mice that we have derived. I already emphasized the error to be attached to our estimate, and the error to be attached to the mouse extrapolation is also considerable. The two values may yet be shown to differ by a factor of 2. On the other hand, the data probably exclude a doubling-dose value for chronic radiation of less than 2 Sv equivalents for either species.

The Sellafield Data

In a recent case-control study attempting to elucidate the cause of a cluster of cases of childhood leukemia in the vicinity of the Sellafield nuclear reprocessing plant located in West Cumbria, United Kingdom, the most significant finding was an association between the occurrence of leukemia and the employment of the affected child's father in the plant (*1*). It was expected that 0.6 of the fathers of the four children with leukemia would be employed in the plant, whereas, in fact, there were 4, yielding a statistically significant relative risk of 6.2. The total estimated external dose of chronic ionizing radiation during employment at Sellafield for these fathers prior to the conception of the children who developed leukemia averaged about 0.15 Sv equivalents; the gonadal dose was perhaps two-thirds of the external dose. The investigators placed a genetic interpretation on the finding. The report has attracted a great deal of popular, scientific, and legal attention over the past several years.

Elsewhere, I have discussed in some detail why the genetic sensitivities to radiation implied by this finding are in flat conflict both with the experimental data on mice and the results of the studies in Japan and will not repeat these arguments here (*28, 29*). Little (*30*), equating film badge exposures to gonadal doses, concludes that the data from Sellafield imply genetic sensitivities 50–80 times greater than the Japanese data. With respect to apparently relevant animal experiments, Gardner and associates (*1*) quote with approval the studies of Nomura (*31*) on the increase in tumors, including leukemia, in the offspring of radiated male mice but neglect to mention that many mouse strains have been selected to exhibit high cancer frequencies, that is, represent a sensitized soil from which extrapolation to a heterogeneous human population may be inappropriate (31).

In Nomura's studies, the frequency of leukemia in the control mice of series 2 was 0.4 ± 0.4%, with an average dose of about 2.5 Gy of ionizing radiation to spermatogonia; this frequency increased to 3.9 ± 1.2%, a 9.8-fold increase (32). With allowance for the factor of 3 to be introduced in extrapolating from acute to chronic effects, the Sellafield data suggest a sensitivity roughly 46-fold greater than the results of these experimental studies on an animal that is probably especially sensitive to this particular effect. While the error to be attached to the foregoing calculation is difficult to assess, the fact that even the Nomura studies do not support the Sellafield findings seems beyond doubt.

The Gardner report has sparked a number of supplementary studies, some still ongoing, but of the two that have come to fruition, one so overlaps with the Gardner study that it cannot be regarded as independent (33) and the other offers no support to the results of the Sellafield study (34). I conclude that either the relative risk obtained by Gardner and associates (1) is an example of the instability of relative risk calculations when expectation is so small that the classification of a single individual can make the difference between a statistically significant and nonsignificant result, or that, if the association is real, it cannot be genetic.

The current policy of the U.S. government, administered by such agencies as the Environmental Protection Agency, is that the citizens of this country as a group should not receive an average accumulated dose of radiation up to about age 30 of more than 0.05 Sv equivalents from nonmedical man-made sources; those who are occupationally exposed should not receive more than 0.25 Sv equivalents. To provide some perspective, all of us on average receive during that 30-year period about 0.05 Sv equivalent from natural sources of radiation— the soil, principally as radon, cosmic rays, radioactivity from the burning of fossil fuels, and the natural radioactivity of our own bodies (25). Various types of monitoring reveal that, in fact, man-made, nonmedical exposures are running well below these permissible limits, the most current estimate being about 0.02 Sv equivalents. From the genetic standpoint, these recommendations, developed some 30–35 years ago, were dominated by the mouse data. In light of what is presented in this chapter, they now seem quite conservative. Even so, no geneticist claims these exposures are genetically completely harmless but only that, considering the benefits received from the processes that generate these exposures, the trade-off seems justifiable.

What Can Be Done To Improve Our Current Understanding of the Importance of Environmental Mutagens?

The issue of the significance of the mutagens in our environment will remain with us indefinitely. With respect to radiation, during the past 6 years, the groundwork has been laid at the Radiation Effects Research Foundation for a major new study directed at detecting mutation at the DNA level. Should that study yield results consistent with the present estimates of the genetic doubling dose of ionizing radiation and should the role of ionizing radiation in the Sellafield leukemias be laid to rest by the additional studies already completed or in progress, then the totality of the studies in Japan will have provided a solid perspective on the genetic risks of ionizing radiation to humans. The situation is much more ambiguous with respect to the chemical mutagens for which there are environmental, occupational, or therapeutic exposures.

It is very possible that public concern will force epidemiological studies of the somatic and genetic effects of some of these exposures within the next decade. It is hoped that, if these studies occur, a positive interaction between the government, the chemical industry, and the scientific community will result in one or more major and definitive studies rather than a series of small studies of low resolving power, where the occurrence of one or two people in a particular data cell is the difference between no effect and a statistically significant finding. My suggestion is that, if public concern were to provoke such studies, the initial major effort should be directed toward a "worst cases" analysis, which in effect is what was done for radiation in the follow-up studies in Japan. At the moment such an analysis would be represented by a careful study of the offspring of individuals treated for childhood malignancies with such potent mutagens as nitrogen mustard, procarbazine, cyclophosphamide, or doxorubicin. The relatively high frequency of secondary tumors in these individuals (35–39) suggests that somatic mutations have been induced; what about germinal mutations? A study pursuing all the end points employed in Japan is indicated, except that studies at the DNA level can now replace the electrophoretic studies. The number of such children is not large, but the unfolding DNA technologies are such that even a single individual and his or her parents can be highly informative. If no germline genetic effect is observed, then the risk of lesser exposures is negligible; if an effect is observed, the first steps are taken toward defining the potential problem of lesser chemical exposures.

In closing I stress that the thrust of this chapter should not be misunderstood. If our hypothesis is correct, the genetic risks of radiation are substantially less than projected. However, this does not mean we should stop worrying, but only worry less. The risks are lower than presumed earlier, and current official guidelines for exposure are quite adequate.

References

1. Gardner, M. J.; Snee, M. P.; Hall, A. J.; Powell, C. A.; Downes, S.; Terrell, J. D. *Br. Med. J.* **1990**, *300*, 423–429.
2. Neel, J. V.; Schull, W. J. *The Effect of Exposure to the Atomic Bombs on Pregnancy Termination in Hiroshima and Nagasaki*; National Academy of Sciences–National Research Council: Washington, DC, 1956; Publ. 461, pp xvi and 241.
3. Kato, H.; Schull, W. J.; Neel, J. V. *Am. J. Hum. Genet.* **1966**, *18*, 339–373.
4. Schull, W. J.; Otake, M.; Neel, J. V. *Science (Washington, D.C.)* **1981**, *213*, 1220–1227.
5. *The Quality Factor in Radiation Protection*; International Commission on Radiation Units and Measurements (ICRUM); Rep. 40; ICRUM: Bethesda, MD, 1986; pp 1–32.
6. Otake, M.; Schull, W. J.; Neel, J. V. *Radiat. Res.* **1990**, *122*, 1–11.
7. Yoshimoto, Y.; Schull, W. J.; Kato, H.; Neel, J. V. *Radiation Effects Research Foundation Technical Report*, 1–91, 1991; p 27.
8. Yoshimoto, Y.; Neel, J. V.; Schull, W. J.; Kato, H.; Mabuchi, K.; Soda, M.; Eto, R. *Am. J. Hum. Genet.* **1990**, *42*, 1053–1072.
9. Awa, A. A.; Honda, T.; Neriishi, S.; Sofuni, T.; Shimba, H.; Ohtaki, K.; Nakano, M.; Kodama, Y. In *Cytogenetics: Basic and Applied Aspects*; Obe, G.; Basler, A., Eds.; Springer-Verlag: Berlin, Germany, 1987; pp 166–183.
10. Neel, J. V.; Satoh, C.; Goriki, K.; Asakawa, J.; Fujita, M.; Takahashi, N.; Kageoka, T.; Hazama, R. *Am. J. Hum. Genet.* **1988**, *42*, 663–676.
11. Schull, W. J.; Neel, J. V.; Hashizume, A. *Am. J. Hum. Genet.* **1966**, *18*, 328–338.
12. Furusho, T.; Otake, M. *Radiation Effects Research Foundation Technical Report*, 4–78, 1978.
13. Furusho, T.; Otake, M. *Radiation Effects Research Foundation Technical Report*, 5–78, 1978.
14. Furusho, T.; Otake, M. *Radiation Effects Research Foundation Technical Report*, 14–79, 1979.
15. Furusho, T.; Otake, M. *Radiation Effects Research Foundation Technical Report*, 1–80, 1980.
16. Furusho, T.; Otake, M. *Radiation Effects Research Foundation Technical Report*, 9–85, 1985.
17. Neel, J. V.; Schull, W. J.; Awa, A. A.; Satoh, C.; Kato, H.; Otake, M.; Yoshimoto, Y. *Am. J. Hum. Genet.* **1990**, *46*, 1053–1072.
18. Sankaranarayanan, K. *Am. J. Hum. Genet.* **1988**, *42*, 651–662.
19. Ehling, U. H. *Annu. Rev. Genet.* **1991**, *25*, 255–280.
20. Russell, W. L. *Cold Spring Harbor Symp. Quant. Biol.* **1951**, *16*, 327–336.
21. Russell, W. L.; Russell, L. B.; Kelly, E. M. *Science (Washington, D.C.)* **1958**, *128*, 1546–1550.

22. Russell, W. L. *Proc. Natl. Acad. Sci. U.S.A.* **1965,** *54,* 1552–1557.

23. *Report of the United National Scientific Committee on the Effects of Atomic Radiation;* United Nations, General Assembly Official Records; Twenty-seventh Session; Supplement No. 25 (A/8725), 1972.

24. *The Effects on Populations of Exposure to Low Levels of Ionizing Radiation: 1980;* Committee on the Biological Effects of Ionizing Radiation, National Research Council; National Academy Press: Washington, DC, 1980; pp xv and 524.

25. *Health Effects of Exposure to Low Levels of Ionizing Radiation (BEIR V);* Committee on the Biological Effects of Ionizing Radiation, National Research Council; National Academy Press: Washington, DC, 1990; pp xiii and 421.

26. Neel, J. V.; Lewis, S. E. *Annu. Rev. Genet.* **1990,** *24,* 327–362.

27. Lüning, K. G.; Searle, A. G. *Mutat. Res.* **1971,** *12,* 291–304.

28. Neel, J. V. *JAMA, J. Am. Med. Assoc.* **1991,** *266,* 698–701.

29. *The Children of Survivors of the Atomic Bombs: A Genetic Study;* Neel, J. V.; Schull, W. J., Eds.; National Academy Press: Washington, DC, 1991; pp vi and 518.

30. Little, M. P. *J. Radiol. Prot.* **1990,** *10,* 185–198.

31. Nomura, T. *Nature (London)* **1982,** *296,* 575–577.

32. Nomura, T. *Genetic Toxicology of Environmental Chemicals Part B: Genetic Effects and Applied Mutagenesis;* Alan R. Liss, Inc.: New York, 1986; pp 13–20.

33. McKinney, P. A.; Alexander, F. E.; Cartwright, R. A.; Parker, L. *Br. Med. J.* **1991,** *302,* 681–687.

34. Urquhart, J. D.; Black, R. J.; Muirhead, M. J.; Sharp, L.; Maxwell, M.; Eden, O. B.; Jones, D. A. *Br. Med. J.* **1991,** *302,* 687–692.

35. Meadows, A. T.; Baum, E.; Fossati-Bellani, F.; Green, D.; Jenkin, R. D. T.; Marsden, B.; Nesbit, M.; Newton, W.; Oberlin, O.; Sallan, S. G.; Siegel, S.; Strong, L. C.; Voûte, P. A. *J. Clin. Oncology* **1985,** *3,* 532–538.

36. Strong, L. C.; Stine, M.; Norsted, T. L. *J. Natl. Cancer Inst.* **1987,** *79,* 1213–1220.

37. Tucker, M. A.; Coleman, C. N.; Cox, R. S.; Varghese, A.; Rosenberg, A. S. *New Engl. J. Med.* **1988,** *318,* 76–81.

38. de Vathaire, F.; Schweisguth, O.; Rodary, C.; François, P.; Sarrazin, D.; Oberlin, O.; Hill, C.; Raquin, M. A.; Dutreix, A.; Flamant, R. *Br. J. Cancer* **1989,** *59,* 448–452.

39. de Vathaire, F., François, P.; Hill, C.; Schweisguth, O.; Rodary, C.; Sarrazin, D.; Oberlin, O.; Beurtheret, C.; Dutreix, A.; Flamant, R. *Br. J. Cancer* **1989,** *59,* 792–796.

RECEIVED for review August 7, 1992. ACCEPTED revised manuscript April 12, 1993.

11

Studies of Children In Utero during Atomic Bomb Detonations

Y. Yoshimoto, M. Soda, William J. Schull,[1] and K. Mabuchi

Radiation Effects Research Foundation, Hiroshima 732, Japan

Although the study of mental retardation among children exposed to atomic radiation in utero has clearly shown an effect of exposure on the developing brain, the cancer risk among these individuals remains to be determined and will only be established through continued follow-up of the subjects. To this end mortality and morbidity surveys of about 1800 persons exposed in utero to the atomic bombings of Hiroshima and Nagasaki were undertaken at the Radiation Effects Research Foundation (RERF). In the years 1950–1984, when these individuals were under the age of 40, a significant excess cancer risk was observed. The relative risk at 1 Gy was about 3.8. However, in the most recent 5 years (1985–1989), there was no apparent excess of cancer, and the overall relative risk for the years 1950–1989 decreased to about two based on the 2 dozen cancer cases with DS86 dose estimates that have been seen thus far.

$\mathbf{A}$ LONG-TERM FOLLOW-UP OF MORTALITY among the atomic bomb (A-bomb) survivors of Hiroshima and Nagasaki has been underway at the Atomic Bomb Casualty Commission (ABCC) and its successor, the Radiation Effects Research Foundation (RERF), since 1950. However, the cohort of survivors on which this surveillance rests, known as the Life Span Study (LSS) sample, does not include individuals exposed prenatally. Since the human fetus is thought to be particularly sensitive to exposure to ionizing radiation, a separate group of in utero exposed persons is also under surveillance (1).

[1]Current address: Epidemiology Research Center, School of Public Health, University of Texas Health Science Center, Houston, TX 77225

As reviewed by Miller (2) there is no doubt that prenatal exposure to atomic bomb radiation induced adverse effects, such as small head size and mental retardation. However, earlier studies of childhood cancer among these survivors failed to reveal a significant increase in mortality attributable to malignancy (3), a finding that seems contradictory to data from the Oxford Childhood Cancer Survey in England that shows an association between diagnostic radiography of pregnant women and childhood cancer risk (4). Although only 13 cancer cases in the ≥ 0.01-Gy-dose atomic bomb survivor group and five in the 0-Gy-dose group were identified during the period 1950–1984, cancer risk appeared to increase significantly as maternal uterine dose increased (5). We hypothesize that this apparent excess cancer risk is caused by exposure to atomic bomb radiation, but to determine more reliably whether the pattern of adult-onset cancer has been altered, follow-up of this cohort is continuing at the RERF.

One of our interests is determining how the cancer risk among the in utero exposed survivors compares with that of survivors who were less than 10 years old at the time of the bombings (ATB). To date, based on the LSS mortality data, the relative risk of cancer at 1 Gy appears to be substantially higher among the latter age ATB group than in any other age group ATB (6).

Pregnant Women ATB

The Hiroshima bomb exploded on Monday, August 6, 1945, and the atomic bomb dropped on Nagasaki on Thursday, August 9. Efforts to reconstruct the probable number of pregnant women exposed to these bombings, based on the distribution of houses in the two cities, are not simple. Hiroshima is built on a fan-shaped, flat delta. However, Nagasaki is more mountainous, and the houses there stood at various levels on the mountainsides. In Table I the number of dwelling houses within 2 km of the hypocenters is calculated assuming that, at the time of the bombing, 51% of the 76,000 buildings in Hiroshima and 25% of the 51,000 buildings in Nagasaki where within this distance (7).

In 1945 most births in Japan occurred at home and were attended by a midwife. [In 1947, for example, in urban areas, only 6.5% of the births were hospitalized deliveries (8)]. The estimated number of pregnant women ATB ranges between 2200 and 4400 for Hiroshima and between 900 and 1800 for Nagasaki (Table I). The lower value is based on the number of survivors known to the ABCC/RERF to be exposed in utero between 2.0–2.9 km from the hypocenters; there are 822 such individuals in Hiroshima and 445 in Nagasaki (1). The upper value is based on the number of live births reported in 1947—67,757 in

Table I. Number of Children In Utero ATB within 2 km
from the Hypocenters Identified by the ABCC/RERF Compared
with the Crudely Estimated Number of Dwelling Houses and the
Expected Number of Pregnant Women

Parameter	Number
Hiroshima	
Dwelling houses	39,000
Pregnant women	2,200
Deaths of mother <20 days	991
Deaths of mother 20–280 days	66
Miscarriages and stillbirths but mother alive	204
Children in utero ATB identified by ABCC/RERF	939
Nagasaki	
Dwelling houses	13,000
Pregnant women	900
Deaths of mother <20 days	635
Deaths of mother 20–280 days	27
Miscarriages and stillbirths but mother alive	59
Children in utero ATB identified by ABCC/RERF	179

Hiroshima Prefecture and 53,021 in Nagasaki Prefecture. The absolute number of live births in 1945, at the end of World War II, is assumed to be 0.85 of that in 1947. Although the atomic bombing was at 8:15 a.m. in Hiroshima, the Nagasaki bombing was closer to noon (11:02 a.m.). Thus, some pregnant women may have been outside of their homes at the time of the bombing, particularly in Nagasaki. However, we assume that the distribution of pregnant women ATB by distance from the hypocenter was similar to that of the general population estimated by the Joint Commission of Japanese and American investigators (9). We ignore the fact that some pregnant women were compulsorily evacuated from these cities prior to August 1945.

Table I provides the numbers of pregnant women estimated to have died within the first 20 days after the bombings and those who died in the following 20–280 days. These are based on the studies by the Joint Commission (9) and Ishida and Matsubayashi (10). We assume that most of the children in utero ATB within 2 km were ascertained by the ABCC/RERF, and they number 939 in Hiroshima and 179 in Nagasaki (1). To support the belief that ascertainment is relatively complete, the number of miscarriages and stillbirths among pregnant women who survived the first 280 days after the bombings was calculated; the rates are assumed to be 0.179 for Hiroshima and 0.248 for Nagasaki. A statistical survey of 20-day survivors (11) and a later study (12) by Yamazaki et al. conducted in 1951 show that the rate of miscarriages and stillbirths within 2 km was about 0.102–0.357. This large range indicates the uncertainties due to incomplete ascertainment of the pregnant women in the studies.

Atomic-Bomb Survivors Exposed In Utero

Survivors exposed in utero are defined as those individuals born to atomic-bomb-exposed pregnant women between the time of the bombings and May 31, 1946. Studies of these survivors were initiated in Nagasaki through the registration of pregnancies associated with the genetic studies (12). In Hiroshima the first survey was conducted in 1950, but it included only those children who were presumably exposed in the first half of their gestational development (13). Over the years a clinical sample has been established in which the observations for severe mental retardation are complete (14). In this sample the number of children exposed in utero within 2 km is 537, about half of the total 1118 (=939 + 179) children determined by ABCC/RERF in Table I.

In 1960 an in utero mortality cohort was established, and a systematic ascertainment of cause of death was begun retrospectively (1). This cohort is the basis for the follow-up studies of cancer incidence as well as mortality. Details concerning the selection of this sample are given elsewhere (5). The results to follow are based on the 1791 individuals, including 161 deaths observed before October 1950, indicated in Table II, because atomic-bomb radiation doses for these individuals were estimated based on the previous dosimetry system (T65D; see the subsequent discussion). This cohort includes most, if not all, of the atomic-bomb survivors exposed in utero within 1.5 km from the two hypocenters. About one-fourth of the 1118 children ex-

Table II. Number of Children Exposed In Utero Including 161 Deaths Observed before October 1950

Parameter	Number[a]	
City		
Hiroshima	1534	(1401)
Nagasaki	257	(229)
Sex		
Male	852	(765)
Female	939	(865)
Data source		
Birth records	1263	(1102)
Others	528	(528)
Trimester ATB		
First	574	(532)
Second	687	(622)
Third	530	(476)
Total	1791	(1630)

[a]The values in parentheses are the number of children who were alive on or after October 1950.

posed in utero within 2 km in Table I are not included, because some were not selected or because of insufficient information on dose estimates. Since 70% of the in utero survivors were identified through birth records, some mortality data for the first 5 years after the bombings are available. The other members of the sample were ascertained through the master file of all survivors registered with ABCC/RERF and the 1960 national census.

The trimesters of pregnancy ATB in Table II are based on the reported birth date and are defined as follows (the numbers in parentheses are for Nagasaki): first trimester: 7(10) February 1946–May 31, 1946; second trimester: 7(10) November 1945–6(9) February 1946; and third trimester: 6(9) August 1945–6(9) November 1945.

Dosimetry and Methods

Radiation-related risks among the atomic-bomb survivors are now analyzed using the Dosimetry System 1986 (DS86) (*15*). Since DS86 fetal-absorbed doses (the actual doses delivered to fetuses) are not yet available, maternal uterine doses are used as a surrogate. DS86 Version 3 dose estimates were computed in 1989, and there were some minor changes in the assigned doses from those used in previous publications (*5*). However, if the analysis is to be extended to all subjects in Table II, ad hoc uterine doses must be assigned to 12.7% of the individuals. These ad hoc doses were computed on the basis of an empirical conversion from the previously assigned Tentative 1965 Doses (T65D and revised T65DR) (*15*, *16*). One thousand and forty-seven children are thought to have been exposed to a maternal uterine dose of 0.01 Gy or more (the mean dose is 0.295 Gy), and the doses of the remaining 772 are 0 Gy. In the comparison to follow of the findings on the prenatally exposed children with those on survivors exposed in the first decade of life, DS86 shielded kerma doses are used for the survivors exposed when less than 10 years old ATB.

The relative risk at 1 Gy of maternal uterine dose (total dose of neutron and gamma) was calculated using an additive relative risk model in a Poisson regression analysis (*17*). The model was used for testing a hypothesis that mortality or cancer risk increases with increasing dose. Deaths, cancer incident cases, and associated person-years were stratified by city, sex, dose (0, 0.01–0.09, 0.10–0.49, 0.50–0.99, ≥1.00 Gy), and attained age (<1, 1–4, 5–14, 15–24, 25–34, ≥35 yr old). Finally, we assume that the large errors are included in the doses of some individuals for convenience.

Results and Discussion

Mortality among Atomic-Bomb Survivors Exposed In Utero.
Deaths in the sample were determined by searching the official Japanese family registries, known as *koseki*, in which each child is registered at birth. Using the koseki, ascertainment of deaths occurring throughout Japan is considered to be 99% complete, and cause of death can be obtained from the death certificates. The first report on mortality among in utero exposed children born alive failed to demonstrate a significant relationship between mortality and distance from the hypocenter (*1*). However, a subsequent report (*18*) using T65D dose estimates found significant increases in mortality with increasing dose among those dying in the first year of life and at or after age 10.

Through December 1989, 237 deaths occurred among the 1791 in utero survivors in Table II, about 43% of these deaths occurring in the first year of life. The relative risk at 1 Gy for all deaths (1945–1989) is higher than one when the data are restricted to the subcohort of 1563 persons with DS86 Version 3 dose estimates (Table III). The difficulty in determining the cause of death for 47 of the children in Table III is partly due to the fact that the systematic confirmation of cause of death did not begin until 1960 and was initially retrospective and partly because the vital statistics records are incomplete for the period immediately following the atomic bombings. All but two of these 47 deaths occurred before October 1950.

A significant increase in mortality from all causes with increasing dose is observed. At face value the relative risk at 1 Gy for fatal cancer seems to be higher than that for other diseases. However, the only specific cause of death found to be related to radiation dose is perinatal death among those 1135 persons who were identified through birth records. Mortality during the first year of life in the ≥1.0 Gy dose group is about three times higher than in the 0 Gy dose group, but 9 of the 11 deaths in the ≥1.0 Gy dose group occurred among survivors exposed in the third trimester, and mechanical injury may have played a role in their deaths. These data shed no light on the hypothesis that individuals whose reticuloendothelial systems were damaged by radiation would presumably be more prone to cancer, resulting in selective elimination (*19*).

In the subcohort on Table III there is one individual, a male, who was previously diagnosed as a cancer incident case of histiocytosis X (*5*) but who died at age 41 in 1986 of Wegener's syndrome (autopsy report). He was exposed in the third trimester ATB to 0.58 Gy maternal uterine dose. At age 6 he received radiotherapy to the spleen

Table III. Updated Number of Deaths and Cancer Cases, 1945–1989, among the Survivors Exposed In Utero When Restricted to the Subcohort of 1563 Persons with DS86 Doses

Parameter	Number
Deaths, total	219
Cause of death	
All diseases	147
Fatal cancer	13(2)
Other diseases	134
External causes	25
Unknown causes	47
Cancer incident cases, total	24(2)
By sex	
Male	6(1)
Female	18(1)
By trimester	
First	4(1)
Second	10
Third	10(1)
By attained age	
<35 year	13(2)
≥35 year	11

NOTE: Number of cancer incident cases include fatal cancer cases. Numbers in parentheses are the numbers of leukemia cases.

for splenomegaly, and 5 years later he had a splenectomy for Banti's syndrome. His experience is unusual, but it is impossible to determine whether his illness was caused by atomic-bomb radiation exposure. This case has been excluded in the following analysis of cancer risk because of the autopsy findings.

The subcohort in Table III includes 17 individuals who were exposed in utero within 1.5 km from the hypocenter based on the DS86 Version 3 dosimetry and are (were) mentally retarded (*14*, *20*). Six of these persons, 35%, died between the age of 5–24 years. This proportion is roughly three times higher than the 10% of deaths before age 25 seen in the 0 Gy dose group. Generally, the life expectancy of individuals with mental retardation is shorter than that of the average individual in the average population (*21*). In addition to these cases 20 individuals among the children within the 1.5 km zone (*22*) were identified as having small head size without mental retardation, but only one death was observed. However, these children had to have survived until the circumference of their head was measured. This occurred in the period from 9 to 19 years of age, that is, after the force of mortality has lessened in children.

Finally, no cancer deaths were observed in the first 5 years of life, that is, before October 1950, although there were many deaths from unknown causes.

Cancer Risk among Atomic-Bomb Survivors Exposed In Utero. Cancers were ascertained through death certificates and through the tumor registries in Hiroshima and Nagasaki. Nonfatal cancers are ascertained in the registry catchment areas only. Ideally, the cancer risks would take into account the effect of migration, but the observed cases are too few to suggest that adjusting the data for migration would make much difference in the findings. Thus, our analysis does not take migration into account. Here we evaluate the cancer risk based on the person-years accumulated in the period 1950–1989, because we want to compare this risk with that of the survivors exposed when less than 10 years old ATB.

In 1950–1989, 24 cancer cases were identified among the subcohort of 1413 in utero exposed atomic-bomb survivors (Table III). For cancer risk in 1950–1984 a significant excess risk was observed, and

Table IVA. Cancer Occurrence of Children Exposed In Utero and Those Exposed at Age <10 ATB

Parameter	In Utero[a]		Age <10 ATB[b]	
	0 Gy	≤0.01 Gy	0 Gy	≤0.01 Gy
Number of subjects	710	920	6901	8994
Number of cancer cases	5(0)[c]	13(2)	49(7)	93(24)

[a]Cancer occurred between 1950 and 1984.
[b]Cancer occurred between 1950 and 1985.
[c]Number of leukemia cases is shown in parentheses.
SOURCE: Adapted from reference 5.

Table IVB. Previously Estimated Cancer Risk of Children Exposed In Utero and Those Exposed at Age <10 ATB

Risk	Exposure In Utero		Exposure at Age <10 ATB	
	Cancer Type	Estimated Values	Cancer Type	Estimated Values
Relative risk at 1 Gy	All cancers	3.77	Leukemia	17.1
			Other Cancers	2.35
Excess risk per 10[4] person-year-Gy	All cancers	6.57	Leukemia	2.93
			Other cancers	2.29

NOTE: Estimated cancer risk is based on an additive risk model in which the differences by city and sex in the background were stratified and the risks determined using a Poisson model.
SOURCE: Adapted from reference 5.

the relative risk at 1 Gy was about 3.8 (5) (Table IV). In the most recent 5 years (1985–1989), no excess of adult-onset cancer is seen, and no new case occurred in the ≥0.50 Gy dose group. No additional leukemia cases were observed either. The overall relative risk decreases to about two when the observation period is extended to the end of 1989. Thus far, more cancer cases have been observed among women than men, but no clear change in the cancer risk estimate is observed even when the analysis is restricted to females only. No evidence exists that the risk of cancer associated with radiation exposure differs by gestational period ATB. The 24 cancer cases are briefly discussed in the following sections.

Cancer in Childhood.   During the first 14 years of life, there were only two cancer cases; both cases were exposed to relatively high doses and were reported by Jablon and Kato (3). The first case was a girl who died of liver cancer at age 6 and who was reported to be mentally retarded. She had been exposed to 1.39 Gy in the first trimester. The second case was a female who developed Wilms' tumor at age 14 and subsequently died of stomach cancer at age 35. She had been exposed to 0.56 Gy in the second trimester.

Leukemia.   Thus far, Ishimaru et al. (23) identified two cases of leukemia among the in-utero-exposed individuals. Both cases were exposed to low doses. One involved acute myelogenous leukemia in an 18-year-old woman (exposed to 0.02 Gy in the third trimester) and the other, acute lymphatic leukemia in a man aged 29 (exposed to 0.04 Gy in the first trimester).

Other Cancers in Adulthood.   Nine cancer cases have occurred in the 0 Gy dose group (four breast, two stomach, and one each of uterus, lung, and metastatic). The age at onset ranged from 29 to 43 years (the mean was 37.6 yr).

In the ≥0.01 Gy dose group, 11 cancer cases were observed (five stomach, two breast, and one each of bladder, choriocarcinoma, thyroid, and ovary). The age at onset ranged from 22 to 42 years (the mean was 32.8 yr). Of these cases, two are in the ≥0.50 Gy dose group, one each of stomach (exposed to 0.90 Gy in the second trimester) and ovarian cancer (exposed to 2.13 Gy in the third trimester). One case previously diagnosed as histiocytosis was deleted in the analysis (see the earlier section on mortality).

Comparison of Cancer Risk between the In Utero Exposed and Survivors Exposed When Less Than 10 Years Old ATB.

ATB.   Two questions have motivated this comparison. First, why does the leukemia risk among the in utero population not increase even though the excess risk of leukemia among atomic-bomb survivors is the typical risk associated with radiation-induced cancers? Second, is the pattern of adult-onset cancer altered among the prenatally exposed survivors? Here, we discuss only issues associated with the first question, since the second cannot be answered with the data currently available, and an answer must await further observations on this group of survivors as they age.

Jablon et al. (24) examined the observed cancer cases for the period 1950–1969 among the less than 10-years-old-ATB group, and Bizzozero et al. (25) examined the leukemia risk for 1946–1964 among the atomic-bomb survivors. Based on these data all of the childhood cancer cases occurring before age 15 among the survivors less than 10 years old ATB in the LSS sample were diagnosed as leukemia (Table V). About 80% of the 14 childhood leukemia cases were estimated to have received a shielded kerma dose of ≥0.50 Gy. No childhood leukemia was observed in the in utero group. Thus, there was no evidence that leukemia risk in childhood had been altered more in the in utero group than the less than 10-years-old-ATB group. Apart from leukemia there was no increased fatal cancer risk before 1965 in either group of survivors (24). For the less than 10-years-old-ATB group, by 1985 the relative risk of leukemia at 1 Gy shielded kerma was about 17, whereas it was about two for fatal cancers except leukemia (6).

Table V. Number of Childhood Cancer Cases (<15 Years Old) among the Survivors Exposed When In Utero or When Less than 10 Years Old (ATB) in the ABCC/RERF LSS Sample

Age ATB (years)	Maternal Uterine or Kerma Dose (Gy)			Total
	0	0.01–0.49	≥0.50	
In utero	0	0	2[a]	2
<1 yr	0	0	1	1
1	0	0	2	2
2	0	1	2	3
3	1	0	3	4
4	1[b]	0	1	2
5	0	0	1	1
8	0	0	1	1

NOTE: The rough values of the sample sizes in the ≥0.50 Gy dose groups were given in the text. No childhood cancer cases were noted for the other ages ATB, that is, 6, 7, and 9 years.
[a]The two cancers observed among the survivors were one case of Wilms' tumor and one case of liver cancer. All other cases of cancer in this table were diagnosed as leukemias.
[b]This dose is based on the previous T65 dose estimates.
SOURCE: ABCC TRs 17–65 and 7–71 and RERF TR 4–88.

There are two cases of leukemia at ages 18 and 29 among the in utero exposed survivors, but both cases were exposed to doses of 0.05 Gy at most.

When the estimation of cancer risk rests on a 1-year birth cohort, such as the survivors exposed in utero, the observations tend to be inherently unstable because of the small sample size. (The rough values of the sample sizes in the ≥0.50 Gy dose groups follow.) Most of the proximally exposed in utero survivors were identified by birth records. In 1950 identification of survivors less than 10 years old ATB was made more difficult by (a) early migration out of Hiroshima and Nagasaki and (b) early death. We know that the mothers (aged 15–50 years ATB; mean age: 28.7 yr) of in-utero-exposed survivors who were <1.5 km from the hypocenter exceed in number those selected for the LSS sample by a factor of 2.03 for Hiroshima and 1.46 for Nagasaki. These figures may be useful in estimating the number of proximally exposed children less than 1 year old ATB, because most children of this age and their mothers were probably at close proximity to one another at the time of the bombing.

The in-utero-age ATB groups receiving a dose of 0.50 Gy or more (maternal uterine or shielded kerma) number about 130–150 survivors, whereas the less-than-1-year-old-exposed group totals about 70. In 1985 the LSS was extended to include more distally (beyond 2000 m) exposed survivors in Nagasaki and is now known as the LSS-E85 (6); this extension did not alter the number of proximally (within 2 km) exposed subjects who are the same ones originally followed up since 1950 (26). The relative risk of cancer varies for each specific site as well as time. Here, we calculated the relative risk at 1 Gy for cancers of all sites for each 1-year birth cohort. Based on unpublished results of cancer risk in 1950–1989, the highest risk appears to occur in the 3–5-years-old age ATB group; the relative risk at 1 Gy is about four. This higher value is influenced partly by the leukemia risk. There is no strong evidence that the risk of leukemia is highest among survivors exposed in the first year of life.

The observation that the radiogenic cancer risk, thus far mainly due to leukemias, is highest in the less than 10-years-old-ATB group and may be related to the peak age observed for acute lymphoblastic leukemia in childhood (27). The major site of hematopoiesis changes as the embryo or fetus develops. Hemoglobin synthesis is initiated in the yolk sac (28), but then erythropoiesis shifts to the liver and finally to the bone marrow. Human blood lymphocyte counts are well above adult levels at birth, rising further during the first year of life and then gradually falling to adult levels by about the early teens (28). Although Greaves' speculations on the etiology of childhood acute lymphoblastic leukemia are general and not specific with respect to radiocarcinogenesis, they are derived from the developmental biology of the immune system (29) and are presumably applicable to exposed individuals as well.

Further Remarks on the Follow-Up of the In Utero Exposed Survivors. Although clear evidence exists showing an effect of radiation on the developing fetal brain, cancer risk among the atomic-bomb survivors exposed in utero remains to be determined and this occurs only through continued follow-up of the subjects. However, it seems unlikely that an excess of leukemia will appear in the remaining lifetime of these survivors, considering the latent pattern of radiogenic leukemia. But, for solid tumors, further careful study will be needed, because these subjects are now entering the cancer-prone ages. An unequivocal excess of solid tumors may not appear until later in life.

The attained height of the survivors exposed in utero, as well as the height of the survivors exposed when less than 10 years old ATB, appears to be lower than that of the controls (30, 31). Given this apparent impairment of growth, it is important that the health status of these individuals continues to be monitored as they approach middle age to determine whether other late effects of exposure to ionizing radiation, aside from mortality or cancer risk, emerge.

References

1. Kato, H.; Keehn, R. J. TR 13–66; ABCC: Hiroshima, Japan, 1966.
2. Miller, R. W. *Health Phys.* **1990**, *59*, 57–61.
3. Jablon, S.; Kato, H. *Lancet* **1970**, *2*, 1000–1003 (ABCC TR 26-70).
4. Mole, R. H. *Br. J. Cancer* **1990**, *62*, 152–168.
5. Yoshimoto, Y.; Kato, H.; Schull, W. J. *Lancet* **1988**, *2*, 665–669 (RERF TR 4-88).
6. Shimizu, Y.; Kato, H.; Schull, W. J. *Radiat. Res.* **1990**, *121*, 121–141 (RERF TR 5-88).
7. The Committee for the Compilation of Materials on Damage Caused by the Atomic Bombs in Hiroshima and Nagasaki. Hiroshima and Nagasaki. *The Physical, Medical, and Social Effects of the Atomic Bombings*; Iwanami Shoten: Tokyo, 1981.
8. Ministry of Health and Welfare. Vital Statistics Japan; Division of Health and Welfare Statistics, Welfare Minister's Secretariat: Tokyo, Japan, 1989; Vol. 1.
9. Oughterson, A. W.; Warren, S. *Medical Effects of the Atomic Bomb in Japan*; McGraw-Hill: New York, 1956.
10. Ishida, M.; Matsubayashi, I. TR 20-61; ABCC: Hiroshima, Japan, 1961.
11. Oughterson, A. W.; LeRoy, G. V.; Liebow, A. A.; Hammond, E. C.; Barnett, H. L.; Rosenbaum, J. D.; Schneider, B. A. *Medical Effects of Atomic Bombs*; Report NP-3041; USAEC, Office of Technical Information, Technical Information Service: Oak Ridge, TN, 1951; Vol. 5.
12. Yamazaki, J. N.; Wright, S. W.; Wright, P. M. *AJDC* **1954**, *87*, 448–463.

13. Plummer, G. W. *Pediatrics* **1952**, *10*, 687–693 (ABCC TR 29-59).
14. Wood, J. W.; Johnson, K. G.; Omori, Y.; Kawamoto, S.; Keehn, R. J. *Am. J. Public Health* **1967**, *57*, 1381–1390 (ABCC TR 10-66).
15. *U.S.–Japan Reassessment of Atomic Bomb Radiation Dosimetry in Hiroshima and Nagasaki*; Roesch, W. C., Ed.; RERF: Hiroshima, Japan, 1987; Vol. 2.
16. Kato, H.; Schull, W. J. *Radiat. Res.* **1982**, *90*, 395–432 (RERF TR 12-80).
17. Preston, D. L.; Kato, H.; Kopecky, K. J.; Fujita, S. *Radiat. Res.* **1987**, *111*, 151–178.
18. Kato, H. *Am. J. Epidemiol.* **1971**, *93*, 435–442 (ABCC TR 23-70).
19. Kneale, G. W.; Stewart, A. M. *Br. J. Cancer* **1978**, *37*, 448–457.
20. Otake, M.; Yoshimaru, H.; Schull, W. J. TR 3-88; RERF: Hiroshima, Japan, 1988.
21. Casadebaig, F.; Quemada, N. *Soc.-Psychiatry-Psychiatr.-Epidemiol.* **1991**, *26*, 78–82.
22. Otake, M.; Schull, W. J. TR 6-92; RERF: Hiroshima, Japan, 1992.
23. Ishimaru, T.; Ichimaru, M.; Mikami, M. TR 11–81; RERF: Hiroshima, Japan, 1981.
24. Jablon, S.; Tachikawa, K.; Belsky, J. L.; Steer, A. *Lancet* **1971**, *1*, 927–931 (ABCC TR 7-71).
25. Bizzozero, O. J.; Johnson, K. G.; Ciocco, A. TR 17-65; ABCC: Hiroshima, Japan, 1965.
26. Jablon, S.; Ishida, M.; Yamasaki, M. TR 15-63; ABCC: Hiroshima, Japan, 1963.
27. Robert W. Miller, National Cancer Institute, Bethesda, MD, personal communication.
28. Wintrobe, M. M. et al. *Clinical Hematology*, 8th ed.; Lea & Febiger: Philadelphia, PA, 1981.
29. Greaves, M. F. *Leukemia* **1988**, *2*, 120–125.
30. Ishimaru, T.; Nakashima, E.; Kawamoto, S. TR 19-84; RERF: Hiroshima, Japan, 1984.
31. Ishimaru, T.; Amano, T.; Kawamoto, S. TR 18-81; RERF: Hiroshima, Japan, 1981.

RECEIVED for review August 7, 1992. ACCEPTED revised manuscript April 29, 1993.

Cancer Risks among Atomic Bomb Survivors

William J. Schull

Epidemiology Research Center, School of Public Health,
University of Texas Health Science Center, Houston, TX 77225

Forty years of study of the life experiences of the survivors of the atomic bombing of Hiroshima and Nagasaki have established a dose-related increased risk involving a variety of cancers. This list includes leukemia and cancers of the breast, colon, esophagus, liver, lung, ovary, stomach, thyroid, and urinary bladder. There is evidence too of an increase in cancers of the salivary glands and skin, and possibly multiple myeloma, although the data on the latter are limited. Risk is clearly a function of the age of the individual at the time of exposure, the young being generally at higher risk. For solid tumors, this increase does not manifest itself until those ages at which cancer normally increases in frequency.

STUDIES OF THE HEALTH EFFECTS OF EXPOSURE to the atomic bombing of Hiroshima and Nagasaki, begun in 1947, still continue. They focus largely on mortality and morbidity in a series of fixed cohorts, two of which are of interest here, namely, the Life Span Study sample, and the Adult Health Study sample (1). Virtually complete mortality surveillance is possible in Japan because of a unique record resource. A system of obligatory household registration has existed since the 1870s (2). All vital events affecting the composition of a family—adoptions, births, deaths, and marriages—must be reported to the office having custody of a family's register. Under the Life Span Study, on a cyclic basis, the register of every individual alive at the end of

0065-2393/95/0243-0147$08.00/0

the last cycle is examined anew to determine whether he or she is still alive. If a person has died since the last cycle, the fact and place of death are recorded in the register. A copy of the death schedule filed at the regional Health Center can be obtained to learn the cause of death. Follow-up is virtually complete; only rarely is an individual lost to the study, generally as a result of migration out of Japan.

Leukemia

The first malignancy to be unequivocally associated with exposure to atomic radiation was leukemia (3, 4). Retrospectively, the frequency of new cases among the survivors reached a peak about 1952 and has declined steadily since then. The excess of cases had not, however, completely disappeared as recently as 1985. This rate suggests a risk period of at least 40 years following exposure (5). Moreover, when incidence is examined in relation to dose, age at the time of the bombing (ATB), and the calendar time of disease onset, it seems that the higher the dose, the greater was the radiation effect in the early period (before October 1955) and the more rapid was the decline in risk in subsequent years. The leukemogenic effect occurred later among individuals who were relatively older ATB, but it still persisted when the last comprehensive review of the leukemia incidence data occurred.

The risks of malignancy among the survivors are customarily couched in terms of exposure to 1 Gy (or 1 Sv in the case of mixed irradiation). This expression is largely a matter of convenience, and the unit could be different. However, if the risk increases in direct proportion to the dose (i.e., linearly, as appears true for all cancers as a group except leukemia), conversion to any other dose is simple. In terms of the older system of classification of leukemias but the new dosimetry and doses to the bone marrow, the excess relative risk of dying from leukemia is 5.21 per Gy, the excess deaths are 2.95 per 10,000 person-year-Gy, and the attributable risk among individuals exposed to 0.01 Gy or more is 58.5% when all ages are combined (5). Here and elsewhere doses are couched in terms of the DS86 Dosimetry System (6). For illustrative simplicity the assertions just made assume that the risk of leukemia increases linearly with dose, but this is not strictly true. Thus extrapolation of these values to doses of less than 1 Gy would overestimate the actual risks to some degree.

Recently the cases of leukemia occurring among members of the Life Span Study were reclassified according to the French–American–British system, and the accumulated information was reexamined (7). This reanalysis clarified some previously puzzling aspects of the data, but it also raised some new questions regarding radiation-related leukemogenesis. For example, it has been recognized for some time that cases of chronic lymphocytic leukemia occurred only in Nagasaki. Reclassification reveals that most of these cases are, in fact, instances of adult T-cell leukemia. Infection with the HTLV-1 virus associated with this form of leukemia is common in areas of Japan's westernmost major island, Kyushu—including Nagasaki—but is relatively rare in the part of Honshu island where Hiroshima is located.

This newer analysis does not support the notion that the time to onset of leukemia is related to age ATB. However, it does further the belief that time to onset is shorter as the dose increases, especially for acute lymphocytic leukemia and chronic myelogenous leukemia. It also suggests that the effect of exposure is more pronounced on the occurrence of these two forms of leukemia than it is on acute myelogenous leukemia, although the frequency of the latter also rises significantly with dose. It further confirms earlier observations that survivors exposed before the age of 16 are more likely to develop acute leukemia (specifically acute lymphatic leukemia) than older survivors, but the latter are more prone to develop chronic myelogenous leukemia.

Death from Cancers Other Than Leukemia

About 1960 some malignant solid tumors were noticed more frequently than expected among the survivors. This increased risk was first mentioned when Tomin Harada and Morihiro Ishida, in a study of the Hiroshima Tumor Registry data, reported that the incidence of lung cancer was significantly higher among those survivors who were exposed within 1500 m of ground zero (8). Subsequent studies, particularly of mortality ascribed to lung cancer among members of the Life Span Study sample, extended this evidence. The carcinogenic effect thus far seen has been most pronounced among individuals 35 or older at the time of exposure. This is not unexpected, however, because the accumulated evidence strongly suggests that radiation-related malignancies increase in frequency at those ages when the specific cancer normally occurs. Leukemia is the only striking exception. As most cancers are diseases of middle and later life, the failure to see an increase in lung cancer among the young may merely reflect the fact that they have not yet reached the ages at which their increased risk will be manifested.

Another malignancy whose relationship to radiation became apparent at about this time is cancer of the breast (9). Subsequent investigations focused on all of the women in the Life Span Study sample, some 63,000. The most recent of these studies (10) found that the distribution of histologic types of mammary cancers does not vary significantly with radiation dose, and that among all women who re-

ceived at least 0.10 Gy, those irradiated before age 20 will experience
the highest rates of breast cancer in subsequent years. This significantly elevated risk is seen even among those females who were exposed before the age of 5. Apparently breast cancer can be induced
by irradiation of stem cells well before breast budding actually begins.

Retrospectively, breast and lung cancer began to increase about
10 years after the bombing, in the period between 1955 and 1960,
but the additional cases were too few at first to make this fact de-
monstrable. However, it is now recognized that there is a delay following exposure before radiation-related cancers become evident. This
time interval between exposure and the clinical manifestation of the
tumor is often called the *latent period*. Why this period should differ
for leukemia and other cancers is not known. However, it is generally
thought to reflect the difference in the rate of cell division, differ-
entiation, and loss of hematopoietic stem cells, on the one hand, and
the cells of nonhematopoietic tissues, on the other.

In a 1959 study (*11*) of diseases of the thyroid, Dorothy Hollingsworth and her associates noted that carcinomas constituted some 7%
of the total number of cases of thyroid disorders they saw, and that
these malignancies were found more commonly among the heavily exposed individuals. Although thyroid cancer is not usually fatal (less
than 10% of individuals with this malignancy die from it), a succession
of studies confirmed the association of this tumor with radiation exposure. The most recent study found thyroid cancer increased in every
exposure group, but the incidence of these cancers is higher in females than in males.

The risks, whether expressed as excess cases or in relative terms,
are significantly higher in individuals who were under the age of 20
years ATB than in survivors who were 20 or more years old when
exposed.

Dosage Effect. These various studies showed that mortality from
cancers of the lung, breast, and stomach increases with increasing dose.
Recently an increase in mortality from cancers of the colon, esophagus, ovary, and urinary bladder appeared (*5*). Tables I and II show
the excess relative risk, the excess deaths, and the attributable risk in
terms of shielded kerma (kinetic energy released in materials) and organ absorbed dose.

Multiple myeloma also increases significantly with radiation dose
(*12*). The suggestion that this might be so appeared as early as 1964,
but several years elapsed before the number of cases was sufficient to
establish the relationship more securely. Multiple myeloma is a malignancy largely confined to older individuals, persons in their 60s or
over. There is also an increase in salivary gland tumors and in thyroid

Table I. Risk of Site-Specific Malignancies on the Basis
of Shielded Kerma

Site of Cancer	Excess Relative Risk per Gy	Excess Deaths per 10^4 PYGy	Attributable Risk (%)
Leukemia	3.97 (2.89, 5.39)	2.30 (1.88, 2.73)	56.6 (46.3, 67.1)
All except leukemia	0.30 (0.23, 0.37)	7.49 (5.90, 9.15)	8.0 (6.3, 9.8)
Esophagus	0.43 (0.09, 0.92)	0.34 (0.08, 0.67)	12.8 (3.0, 25.0)
Stomach	0.23 (0.13, 0.34)	2.09 (1.20, 3.06)	6.4 (3.6, 9.3)
Colon	0.56 (0.25, 0.99)	0.56 (0.26, 0.91)	15.2 (7.0, 24.9)
Lung	0.46 (0.25, 0.72)	1.26 (0.70, 1.89)	11.6 (6.5, 17.4)
Breast	1.02 (0.48, 1.76)	1.04 (0.53, 1.61)	22.4 (11.5, 35.0)
Ovary	0.80 (0.14, 1.85)	0.45 (0.09, 0.89)	18.6 (3.6, 37.1)
Bladder	1.06 (0.46, 1.09)	0.56 (0.27, 0.90)	23.4 (11.2, 37.4)
Multiple myeloma	1.89 (0.56, 4.45)	0.22 (0.08, 0.39)	32.9 (11.5, 59.8)

NOTE: Estimates are based upon mortality in the Life Span Study sample in the years 1950 to 1985. These estimates are based upon shielded kerma, a linear response model over the whole dose range, and only those members of the sample on whom DS86 doses exist (both cities, sexes, and all ages ATB combined). This table has been adapted from Table 6 in reference 5. Numbers in parentheses indicate the 90% confidence interval.

Table II. Risk of Site-Specific Malignancies Correlated to Organ Dosages

Site of Cancer	Excess Relative Risk per Gy	Excess Deaths per 10^4 PYGy	Attributable Risk (%)
Leukemia	5.21 (3.83, 7.14)	2.94 (2.43, 3.49)	58.6 (48.4, 69.5)
All except leukemia	0.41 (0.32, 0.51)	10.13 (7.96, 12.44)	8.1 (6.4, 10.0)
Esophagus	0.58 (0.13, 1.24)	0.45 (0.10, 0.88)	13.0 (3.0, 25.5)
Stomach	0.27 (0.14, 0.43)	2.42 (1.26, 3.72)	5.7 (3.0, 8.7)
Colon	0.85 (0.39, 1.45)	0.81 (0.40, 1.30)	16.3 (8.0, 26.2)
Lung	0.63 (0.35, 0.97)	1.68 (0.97, 2.49)	12.3 (7.2, 18.3)
Breast	1.19 (0.56, 2.09)	1.20 (0.61, 1.91)	22.1 (11.3, 35.0)
Ovary	1.33 (0.37, 2.86)	0.71 (0.22, 1.32)	22.3 (6.9, 41.4)
Bladder	1.27 (0.53, 2.37)	0.66 (0.31, 1.12)	21.5 (9.8, 35.7)
Multiple myeloma	2.29 (0.67, 5.31)	0.26 (0.09, 0.47)	31.8 (11.0, 57.6)

NOTE: Estimates are based upon mortality in the Life Span Study sample in the years 1950 to 1985. These estimates are based upon organ doses, a linear response model over the whole dose range, and only those members of the sample on whom DS86 doses exist (both cities, sexes, and all ages ATB combined). This table has been adapted from Table 7 in reference 5. Numbers in parentheses indicate the 90% confidence interval.

tumors. An increase in mortality attributable to lymphoma remains uncertain.

No increase has been seen in cancers of the bone, gallbladder, nose and larynx, pancreas, pharynx, prostate, rectum, skin (except melanoma), small intestine, and the uterus. Present mortality evidence also fails to suggest an increase in brain tumors, and it is equivocal with regard to tumors of the central nervous system other than the brain. Whether an increase in deaths ascribed to cancer of the liver occurs is unclear. When the analysis is restricted to those cancers known as primary, liver cancers do not increase significantly with dose; however, if the cancers termed "unspecified" are included, there is a dose-related increase (5). The liver is a common site of metastasis for cancers arising elsewhere (in the breast or lung, for example). The unspecified tumors might be metastatic ones that should be assigned to other organs where an effect of radiation is known to occur. A later analysis based on the tissue and tumor registries suggests that primary cancers of the liver increase in a dose-related manner among the survivors (13). To avoid the poor accuracy of diagnoses of liver cancer on death certificates, this analysis focused on histologically confirmed cases where the excess relative risk at 1 Sv was found to be 0.66 (confidence interval [CI]: 0.11; 1.14).

Effect of Age and Location.   The increase in mortality from cancers other than leukemia becomes significant, generally, when individuals reach the usual age of onset for a given cancer (5). In addition, the distribution of time from radiation exposure to death does not differ significantly by radiation dose for solid tumors, but it does vary according to the age of the individual ATB. Both the relative and absolute risks for cancers other than leukemia are higher for younger ATB cohorts at the same attained age. Among individuals over the age of 20 when exposed, and certainly over the age of 30, the relative risk has changed little with time, although the absolute risk has continued to rise. In the two youngest groups of survivors—those individuals who were 0–9 or 10–19 years old ATB—the relative risk has been declining, significantly so among those 0–9, whereas the absolute risk has steadily increased (5). These findings are not inconsistent statistically because if the relative risk is declining with age while the baseline rate is increasing (as it does with age), even a smaller relative risk applied to a larger baseline may produce a larger absolute risk.

Earlier studies suggested significant differences in the frequency of death attributed to cancer in the two cities following exposure to the same amount of radiation (14). These differences were thought to be due to the far greater exposure to neutrons in Hiroshima than in Nagasaki. Experimental work suggests that an absorbed dose of neutrons is appreciably more carcinogenic than the same absorbed dose of gamma rays. Analyses using the newer individual doses show these city differences to be no longer statistically significant. However, at the same dose, mortality remains generally higher in Hiroshima than in Nagasaki. This fact, coupled with similar findings for the occurrence of epilation and chromosomal abnormalities, suggests that city differences might still exist (*15–17*).

What do these measures of risk and statements about cancer mortality mean to members of the Life Span Study sample, and to the survivors generally? In the years from 1950 to 1985, 202 individuals in the Life Span Study who were assigned DS86 doses died of leukemia. Of these individuals, 141 were exposed to 0.01 Gy or more, and 83 (slightly less than 59% of these deaths) were attributable to radiation exposure. These same years saw 5734 deaths from cancers other than leukemia among the members of the Life Span Study sample, and 3172 of these deaths involved survivors who had been exposed to 0.01 Gy or more. Approximately 8%, or 254, of these 3172 deaths presumably stemmed from radiation exposure. These numbers are merely estimates, and they must be interpreted in this light because it is presently impossible to separate a radiation-related cancer from one resulting from some other cause. Even so, there is clearly no epidemic of cancer deaths among the exposed population. Most survivors who will die of cancer will do so as a result of exposure to other factors (e.g., tobacco smoking or alcohol consumption) and not from their exposure to atomic radiation.

Incidence of Solid Tumors

Estimates of cancer risk based on the tumor and tissue registry data (and hence on incidence rather than death) have not been generally available in the past, although these data have been used in the study of specific tumor sites, such as the breast and thyroid. This situation is changing. Recently, for example, a comprehensive assessment was published (*13*) of solid tumor incidence in Hiroshima and Nagasaki in the years from 1958, the inception of the registries, through 1987. Some 8613 cancers were identified among members of the Life Span Study sample and enrolled in the registries during these years. This total shows 3080 more cancers than were identified through death certificates in the same years and sample. Much of this difference involves cancers of the stomach (795 more cases), the breast (386), the thyroid (182), the skin (152), and the uterus (347), but other sites contribute as well. Some of the discrepancy undoubtedly results from underreporting of cancer on the death certificates and thus reflects the limitations of death certificate data. Other differences arise from the registration of a malignancy that, although not yet fatal, will in time be so.

Broadly speaking, the incidence data support most of the findings in a study of the death certificates. For example, neither set of data reveals a significant difference between the cities in the estimates of radiation-related cancer risk, both suggest that the best fitting dose–response model is a linear one, and both reveal the risk to be higher among those survivors exposed early in life (before the age of 20) than at later ages, although the difference declines with time.

There are differences, however, some expected and some not. Thus, the incidence data generally lead to risk estimates with smaller errors, as would be expected, because the number of incident tumors seen is greater than the number of deaths caused by malignancy at the same site. These data also show a significant increase in thyroid tumors and nonmelanomatous tumors of the skin with increasing dose; this increase has not been clearly seen in the mortality data, presumably because these tumors are not commonly the cause of death. But the incidence data do demonstrate a significant increase in primary cancers of the liver, a finding that has been equivocal in analyses of the mortality data.

Although the estimates of the excess relative risk at 1 Sv do not differ greatly, the attributable risk, based on the tumor registry data, is often higher than that calculated on the basis of the death certificates. The difference is particularly noticeable for organs such as the breast, for which the incidence data suggest an attributable risk of 32% as contrasted with about 22% for the mortality data. Overall, the attributable risk is about 12% when calculated with the registry data and about 8% with the mortality findings. That the former should be higher than the latter is not unexpected because the mortality findings include deaths from cancer in the years 1950 to 1958, before the beginning of the registries, when the contribution of radiation-related cancers to all cancers seen was small.

Uncertainties in the Estimates of Cancer Risk

Many fundamental uncertainties surround these estimates of the carcinogenic effects of radiation. They are both general, in that they are common to all studies of the carcinogenic effects of ionizing radiation, and specific, in that they apply only to the investigations of the survivors. Among the general uncertainties is the absence of a compelling biological model of the underlying process involved not only in radiation carcinogenesis, but in carcinogenesis more broadly. Insofar as ionizing radiation is concerned, the need for a good theoretical model is greatest where the data are weakest—at low doses and low dose rates. At this level our ability to estimate the risk is severely limited and will undoubtedly continue to be so because the excess risk is apparently small and the sample size that would be needed to demonstrate an effect is prohibitively large.

Risk Factors.   Most cancers appear to be environmental in the sense that exposure to environmental factors is involved, but such exposures often embrace a multiplicity of different agents and the nature of their interaction is unknown. In addition, many host factors (such as a person's genes, gender, age, developmental stage, or hormonal status at the time of exposure) affect risk. In any population exposed to ionizing radiation, variation is expected in exposure to these other risk factors as well. At low levels of radiation exposure the effect of this variation might be greater, perhaps much greater, than the risk produced by the radiation exposure itself. It is not surprising, therefore, that it is difficult to estimate risk at low doses or that differing results often occur. The methods used to estimate risk tacitly assume that all exposed individuals in a given category (e.g., age, gender, or dose) have equal risk, which seems unlikely to be true.

The importance of these extraneous modifiers of radiation risk was well demonstrated in the case of breast cancer. Charles Land and his associates detailed the role of reproductive factors in altering the risk of this malignancy in Hiroshima and Nagasaki (personal communication). They showed, for example, that a woman exposed to 2 Gy of atomic radiation in the first decade of life, who has her first child at the age of 18, has a risk of breast cancer that is only one-sixteenth that of a similarly exposed woman whose first child is born when she is 32. Presumably the observed differences in risk are ascribable to hormonal changes subsequent to the woman's exposure, but precisely how these changes and radiation interact is unknown.

Another important risk factor is tobacco smoking. Based on the experience of the survivors, and specifically with respect to cancer of the lung, smoking does not multiply the effect of exposure to irradiation, but merely adds to it (5). Whether this will be true for other sites of cancer that are related to smoking (for instance, cancer of the urinary bladder) is still not known. However, the relationship between smoking, dose, and lung cancer that is seen among the survivors may not apply to other exposures to irradiation such as uranium mining, in which particles are actually deposited in the lung and serve as foci of irritation.

Problems in Epidemiological Studies.   Most of the other uncertainties that attend the estimates of risk resulting from the mortality surveillance in Hiroshima and Nagasaki are not unique to the atomic-bomb data. They are common to all epidemiological studies of the effects of ionizing radiation in humans. Some relate to the doses assigned to specific survivors, but others are more general. We will focus on the latter group because the survivor-specific uncertainties, dependent as they are upon an individual's ability to recollect precisely the details of his or her exposure—where he or she was at the time of the bombing, the presence of shielding, and the like—will never be fully resolved.

Unreconciled differences between the two cities remain. These differences include not only the increased mortality seen in Hiroshima as contrasted with Nagasaki for the same presumed dose, but also the frequency of epilation, chromosomal aberrations, and lenticular opacities. Whether these differences imply residual inaccuracies in the dosimetric system itself or still unrecognized extraneous sources of variation has yet to be determined. Nonetheless, their existence has prompted some controversy about the estimated yield of the Hiroshima weapon, as well as the neutron flux itself.

Although a linear relative risk model is a simple, suitable descriptor of the actual observations to date on cancers other than leukemia, it is unclear whether an alternative dose–response model would be better. Generally, we are unable to discriminate between it and other plausible alternatives, such as a linear–quadratic model. All of these models are, however, merely convenient descriptions of what is observed and might have no deep causal meaning. Radiobiological considerations could suggest a particular dose–response relationship based upon cellular or molecular events, but it does not follow from this pattern that the same dose–response will be seen when measured in terms of case occurrences or relative (absolute) risk of death. The prospects of early clarification of the "true" dose–response relationship seem poor. Presumably as a larger and larger proportion of the lifetime experience of the atomic bomb survivors accrues, some clarification will occur. (Only 28,737 (38%) of the 75,991 members of the Life Span Study sample included in the last mortality analysis, spanning the years 1950–1985, were dead; the number of deceased continues to increase and had reached 42% in 1990.) However, it is debatable whether the appropriate model can ever be defined solely on the basis of epidemiological data.

Radiation Dose and General Health. The Life Span Study sample represents a selected group, conditioned by the changing probability of survival as a result of changing dose. At high doses relatively few individuals survived, whereas at low doses most did. These facts have a number of implications. It has been argued, largely on statistical grounds, that as a consequence of these differences in survival, doses may be overestimated at the higher levels and underestimated at the lower levels (18). This situation could lead to an underestimation of the risk, which would obscure the true dose–response relationship (19–21; see also references 22 and 23).

Other investigators have contended that exposure resulted in a compromising of the immune system. Consequently, individuals who might eventually have succumbed to cancer died instead of infectious or other diseases in which immune competence is important (*see*, for example, reference 24). It is difficult to test this thesis rigorously because so little is known about the causes of death in the first 9 months following the bombings. Nonetheless, the pattern of mortality has certainly changed with time, and the risk of death of a survivor in the surveillance sample who is now 50 years old is not the same as that of one who achieved his or her 50th birthday 2 decades ago.

This potential selection of unusually healthy individuals is often called the *healthy worker effect*. In occupational cohorts, workers are often healthier than the general population of which they are members. Thus their baseline cancer rates might differ from population rates and complicate the estimation of the expected number of cases in the exposed cohort. Although the possibility of such selection in the Life Span Study cannot be peremptorily rejected, the healthy worker effect is most pronounced in the early years of an investigation and wanes with time. Thus if there were an effect on the data from the Life Span Study it would involve leukemia primarily, because the onset of this malignancy occurred early. In contrast, it would have little impact on the risk of solid tumors, for which a significant increase did not occur until 10 or more years after the bombings.

Tumor and Tissue Registries. Although tumor registries were established in Hiroshima and Nagasaki more than 30 years ago and tissue registries have existed for almost 2 decades, the available risk estimates are based mainly upon mortality. Thus they might underestimate the risk at specific sites, as they certainly must for those cancers that are not commonly fatal. However, about 4900 of the 16,000 or so Life Span Study subjects who died between 1961 and 1975 were autopsied (25). These postmortem examinations served as the bases for a succession of evaluations of the reliability of death certificate diagnoses of cancer and other causes of death. As a rule autopsies have confirmed the cancers reported on death certificates. Confirmation rates vary with cancer site, gender, the age of the individual at death, and whether death occurred at home or in a hospital. However, the confirmation and detection rates appear essentially independent of radiation dose.

The confirmation rate represents the frequency with which an autopsy verifies the cause of death stated on the death certificate, whereas the detection rate is the actual frequency of a specific cause of death as revealed by autopsy. Most studies of the reliability of death certificates, in Japan and elsewhere, have shown that when the death certificate states that cancer was the cause of death, it is usually correct. However, these studies have also shown that possibly one out of every four cancers (all sites combined) goes unreported on the death certificate. Among those frequently not recorded are malignancies of the prostate, the thyroid, and those other organs that are rarely the actual cause of death. Some commonly fatal cancers go unrecognized too. For example, only about one out of every six malignancies of the gallbladder or bile ducts was detected in the years of the autopsy program, and no more than one in five cancers of the cervix of the uterus was diagnosed.

The detection rate can range from as low as 15% for cancers of the liver, gallbladder, and bile ducts to as high as 78% for cancer of the breast. Detection rates do not vary significantly with dose, but may vary with age at occurrence of the cancer, declining as age increases. Underreporting will not affect the estimation of the relative risk importantly if underreporting itself is merely a reflection of the general standards of medical diagnosis and is not related to dose. But it would affect the estimation of the number of excess deaths.

Even the estimates that are now becoming available through the tissue and tumor registries have their limitations because the registries are geographically based and cannot provide incidence cases on the full Life Span Study sample. As yet no national registry exists to supplement local data and provide information on survivors who have migrated to other areas of Japan since the establishment of the Life Span Study. To some extent these limitations can be offset by appropriate statistical techniques (26), and a judicious use of the registry and mortality information should provide a more balanced perspective on risk than the one we have had.

Acknowledgments

The findings described here represent the work of many individuals, but in particular of Yukiko Shimizu, Hiroo Kato, and Dale Preston. Any inadvertent errors of fact or emphasis are, however, my own.

The Radiation Effects Research Foundation (formerly ABCC) was established in April 1975 as a private nonprofit Japanese Foundation, supported equally by the Government of Japan through the Ministry of Health and Welfare and the Government of the United States through the National Academy of Sciences, under contract with the Department of Energy.

References

1. Beebe, G. W.; Usagawa, M. Technical Report No. TR 12–68, 1968; ABCC: Hiroshima, Japan.
2. Naruge, T. *Koseki no Jitsumu to sono Riron;* Nihon Kajo-Shuppan: Tokyo, Japan, 1956; p 620.

3. Folley, J. H.; Borges, W.; Yamawaki, T. *Am. J. Med.* **1952**, *13*, 11–21.
4. Moloney, W. C.; Kastenbaum, M. A. *Science (Washington, D.C.)* **1954**, *121*, 308–309.
5. Shimizu, Y.; Kato, H.; Schull, W. J. Technical Report No. TR 5–88, 1988; RERF: Hiroshima, Japan.
6. *U.S.–Japan Joint Reassessment of Atomic Bomb Radiation Dosimetry in Hiroshima and Nagasaki: Final Report*; Roesch, W. C., Ed.; The Radiation Effects Research Foundation: Hiroshima, Japan, 1987; Vol. 1, p 434.
7. Tomonaga, M.; Matsuo, T.; Carter, R.; Bennett, J. M.; Kuriyama, K.; Imanaka, F.; Kusumi, S.; Mabuchi, K.; Kuramoto, A.; Kamada, N.; Ichimaru, M.; Pisciotta, A. V.; Finch, S. C. Technical Report No. TR 9-91, 1991; RERF: Hiroshima, Japan.
8. Harada, T.; Ishida, M. *J. Nat. Cancer Inst.* **1960**, *25*, 1253–1264.
9. Wanebo, C. K.; Johnson, K. G.; Sato, K.; Thorslund, T. W. Technical Report No. TR 13–67, 1967; ABCC: Hiroshima, Japan.
10. Tokunaga, M.; Land, C. E.; Tokuoka, S. *J. Radiat. Res. Suppl.* **1991**, *32*, 201–211.
11. Hollingsworth, D. R.; Hamilton, H. B.; Tamagaki, H.; Beebe, G. W. *Medicine* **1963**, *42*, 47–71.
12. Ichimaru, M.; Ishimaru, T.; Mikami, M.; Matsunaga, M. Technical Report No. TR 9–79, 1979; RERF: Hiroshima, Japan.
13. Thompson, D.; Mabuchi, K.; Ron, E.; Soda, M.; Tokunaga, M.; Ochikubo, S.; Sugimoto, S.; Ikeda, T.; Terasaki, M.; Izumi, S.; Preston, D. Technical Report in draft; RERF: Hiroshima, Japan.
14. Kato, H.; Schull, W. *J. Radiat. Res.* **1982**, *90*, 395–432.
15. Awa, A. A. In *Chromosome Aberrations: Basic and Applied Aspects*; Obe, G.; Natarajan, A. T., Eds.; Springer-Verlag: Berlin, Germany, 1989; pp 180–190.
16. Preston, D. L.; McConney, M. E.; Awa, A. A.; Ohtaki, K.; Itoh, M.; Honda, T. Technical Report No. TR 7–88, 1988; RERF: Hiroshima, Japan.
17. Stram, D. O.; Mizuno, S. *Radiat. Res.* **1989**, *117*, 93–113.
18. Jablon, S. Technical Report No. TR 23–71, 1971; ABCC: Hiroshima, Japan.
19. Gilbert, E. S. *Radiat. Res.* **1984**, *98*, 591–605.
20. Gilbert, E. S.; Ohara, J. L. *Radiat. Res.* **1984**, *100*, 124–138.
21. Pierce, D. A.; Stram, D. O.; Vaeth, M. Technical Report No. TR 2–89, 1989; RERF: Hiroshima, Japan.
22. Pierce, D. A.; Vaeth, M. In *Low Dose Radiation: Biological Bases of Risk Estimation*; Baverstock, K. F.; Stather, J. W., Eds.; Taylor and Francis: London, 1989; pp 54–69.
23. Preston, D. L.; Pierce, D. A. Technical Report No. TR 9–87, 1987; RERF: Hiroshima, Japan.
24. Stewart, A. M.; Kneale, G. W. *J. Epid. Comm. Health* **1984**, *38*, 108–112.
25. Yamamoto, T.; Moriyama, I. M.; Asano, M.; Guralnick, L. Technical Report No. TR 18–78, 1978; RERF: Hiroshima, Japan.
26. Sposto, R.; Preston, D. L. Commentary and Review, 1992; RERF: Hiroshima, Japan.

RECEIVED for review August 7, 1992. ACCEPTED revised manuscript March 25, 1993.

A Health Assessment of the Chernobyl Nuclear Power Plant Accident

Fred A. Mettler, Jr., and Jonathan E. Briggs

Department of Radiology, School of Medicine, University of New Mexico, Albuquerque, NM 87131–5336

In 1989 the then Soviet government requested that the International Atomic Energy Agency (IAEA) assess the steps it took to protect the health of villagers in areas surrounding the site of the 1986 Chernobyl nuclear power plant accident. The International Chernobyl Project (ICP) performed the assessment. "Task 4" of the ICP studied sample populations from three Soviet republics. Teams of physicians from several nations visited seven "control" (uncontaminated) and six "contaminated" villages to obtain in-depth medical histories on and to perform extensive physical examinations of over 1300 persons. No adverse health effects directly attributable to radiation were found by Task 4. Many of the villagers demonstrated increased stress and anxiety related to the accident, but no significant differences were seen between residents of the contaminated and the control villages. However, a high incidence of hypertension, poor dental health, and obesity in the population samples from all the villages did exist. Although it was too early to see increases in leukemia and solid tumors in the populations examined, the authors expect that there will be increases in the incidence of both these types of cancers over the next several decades.

The Accident

The release of radioactive and other materials from the Chernobyl nuclear power plant began at approximately 1:23 a.m. on Saturday, April

0065–2393/95/0243–0161$08.00/0

26, 1986, just after the roof was blown off of the Unit 4 building. Smoke, fumes, vaporized elements, and debris continued to rise out of the reactor fire for the next several days. The plume that rose from the fire reached a maximum height of 1800 m, decreasing during the first 6 days but increasing from the seventh to the tenth day. Surface winds were light but variable during the release, and there were heavy rains on some days.

Radioactive iodine and cesium were carried to the greatest heights and detected outside Russia. Heavier elements, such as cerium, zirconium, neptunium, plutonium, and strontium, were deposited in significant amounts only within Russia. Estimates of the amount of radioactivity freed from the reactor core are about 50 million curies (1.9 $\times$ 10^{18} Bq), including nearly 10 million curies (37.0 $\times$ 10^{16} Bq) of iodine and 2 million curies (74 $\times$ 10^{16} Bq) of cesium.

Because of the magnitude of the accident, the duration of the release, the fire and resulting plume, and the weather, hundreds of thousands of people were exposed to a variety of radioactive materials through many and varied pathways. Several exposed population groups can be identified. First, workers at the plant at the time of the accident received high doses from lengthy exposure to the open reactor, contaminated water, and fallout. About 200–300 workers suffered from beta burns and acute radiation syndrome, and about 30 of these died. After the fire was extinguished, about 650,000 workers assisted in the cleanup of the plant and construction of the sarcophagus. These so-called "liquidators" were exposed to dose rates that were high but were limited by time. Finally, villagers in thousands of settlements and towns surrounding the plant received low but continuous exposure from fallout and contaminated food.

Little information is available as to how much exposure most of the people in the first two groups received. Consequently, what health effects were and will be attributable to the accident will remain difficult to assess for these people. However, we may assess to a limited degree the medical status of some of the villagers living in settlements around the plant. This information is useful not only in determining the current health of these people, but also as a baseline for future assessments.

Medical Assessment by Task 4 of the International Chernobyl Project

The actual and potential medical effects of the Chernobyl accident greatly concern those people directly involved, former Soviet and present Confederation of Independent States (CIS) authorities, and people all over the world. Soon after the accident stories of increased illness of all types were reported in the press, although it was very difficult to verify these confusing and often conflicting reports. The complexity and scope of the accident made it difficult for the authorities and politicians to determine if what they did and were still doing to protect the people living in areas contaminated from fallout was appropriate and effective.

In late 1989 representatives of the former Soviet government requested the International Atomic Energy Agency (IAEA) to assess the steps it took to protect settlement residents (*1*). The IAEA proposed that experts and consultants from a large number of countries should perform a radiological evaluation, and the International Chernobyl Project (ICP) was born. Participating agencies included The Commission of the European Communities, The Food and Agriculture Organization of the United Nations, The International Labour Office, the United Nations Scientific Committee on the Effects of Atomic Radiation, the World Health Organization, the World Meteorological Organization, and representatives from more than a dozen nations. One of the five parts or "tasks" of the project was to evaluate the health effects from radiation exposure and health in general. "Task 4", as the health evaluation of the general population was called, conducted its assessment in three affected areas: the Ukraine, Byleorus, and the Russian Soviet Federated Socialist Republic.

The Scope of Task 4. The health evaluation focused on four basic questions.

1. What was the current health status of the general population?
2. What health problems were related to the Chernobyl accident?
3. What health effects were directly caused by radiation exposure?
4. What health effects may be expected in the future?

Task 4 leaders used a two-point attack to answer these questions. First, there was a review of Soviet data that was gathered since the accident. These data were analyzed using standard epidemiological criteria. Second, medical examinations were performed of persons from the three republics.

To obtain the existing Soviet data, project physicians met with over 70 scientists in Moscow, Kiev, and Minsk. These data and meetings established the goals of the field studies. A review of Soviet data on nutrition in the area studied was also performed. This review was followed by an independent, though limited, evaluation of the participating settlements' nutritional statuses.

The selection of settlements to be studied was made by the Task 4 leaders with the intent to include communities that were representative of the study areas, especially in terms of socioeconomic factors. Thirteen settlements ranging in population from 3000 to 15,000 were chosen. Of these, seven were "control" settlements and six were "contaminated" settlements. Control settlements were defined as having ground level contamination of less than 1 Ci/km^2 (37 kBq/m^2), and contaminated settlements were defined as having more than 15 Ci/km^2 (555 kBq/m^2) of cesium.

Because time and resources limited the number of persons who could be examined, a representative sample from each settlement was selected. To create such a sample the individuals who underwent examination in the small settlements were selected by the year of their birth, and those in the larger settlements, by the month and year of their birth. Samples from five age groups (2, 5, 10, 40, and 60 years old) numbering about 20 people each were used. This typically resulted in 10–80% of a settlement's population being represented, depending on the settlement's size. Approximately 250 persons from each settlement were examined, totaling 1356 for the overall study. The examinations were conducted in 1990, and thus the 2-year-old children examined had not been born at the time of the accident and the 5-year-old children were under a year old when the accident occurred.

This methodology has some limitations: only those people still residing in the areas selected were examined (it was not possible to identify, locate, and examine those who had left); only small- and medium-sized settlements were visited, and thus urban areas with relatively minor contamination were not represented; and, finally, official data on health prior to the accident were very limited.

Physical Examinations and Data Gathering.   The field teams included specialists from the following disciplines: radiation effects, pediatrics, hematology, thyroid disease, ultrasonography, and internal medicine. A World Health Organization representative also was present on each trip, and on one trip a specialist in psychological disorders was also included. The examinations focused on the following areas: the person's past medical history, general psychological state, general health, cardiovascular status, growth parameters, nutrition, thyroid structure and function, and hematological status, as well as the presence of cataracts or neoplasms. An assessment of biological dosimetry was also made from the blood samples taken.

The medical examinations were performed during three 2-week field trips. In all nearly 3000 people were examined. Of these more than half were self-referred. If, after examination, a disorder was felt to be present, the person was referred to their local health care providers for treatment. Only the findings from the examinations of those selected by age sampling as described previously were included in the project results.

The examinations were quite thorough. First, specially trained interpreters assisted the settlement residents who had been selected in filling out an extensive questionnaire consisting of more than 100 inquiries. Data, such as height, weight, blood pressure, and pulse, were measured, and information about the person's diet was recorded. Then a Task 4 physician performed a physical examination, including a review of the person's medical history, again assisted by an interpreter. Thyroid ultrasonography was done, and thermal images were made of all examinations. Blood samples were taken. These samples underwent a variety of analyses to establish levels of thyroid hormones, hemoglobin, leukocytes, lymphocytes, and levels of potentially toxic materials, such as lead, cadmium, mercury, and boron. (Food samples were also examined for these contaminants.)

Results

Review of nutritional and medical data from Soviet institutions and individual investigators is, in most cases, inconclusive at best. When specific areas of interest were looked at, such as hematology, often a conclusive determination could not be made from the Soviet data as to whether abnormalities had occurred before or since the accident. For example, although levels of hemoglobin were reported to be low in some children from contaminated settlements, there were no data from these children nor were there data for children from comparable but uncontaminated villages. Methodologies for evaluating conditions varied greatly, and terminology was not standardized. In cases where comparisons could be made, the Soviet data studies rarely used controls or standards so that the significance was limited.

The Soviet data did not show a significant increase in the incidence of leukemia or cancers. However, because the Soviet system of typing tumors used categories larger than those of the international system, it was not possible to determine if there were increases in the number of some rare types of tumors nor was it possible to rule out an unseen increase. As for the future it is unlikely that hereditary effects or increases of cancers above the natural incidence can be ascertained using dose estimates and current radiation risk estimates, even if long-term studies are carefully performed.

General Health. No disorders affecting the general health of any of the individuals examined or of any of the population samples taken as a whole could be directly attributed to the Chernobyl accident. It was determined, however, that 10–15% of the adults examined needed to see a doctor for existing conditions, such as hypertension, obesity, or dental problems. Children were generally found to be healthy and growing in accordance with both Russian and current U.S. standards, despite differences in nutrition and health care.

The only generalized finding that could be interpreted as a result of the accident was the high levels of stress and anxiety seen in the settlement residents. This anxiety was most often produced by individual concerns about the future, the prospect of having to relocate, or both. These conditions were much more prevalent than the biological threat of the contamination would warrant. Many of the people examined believed or suspected they had physical problems from the contamination. Relocation also induced anxiety in many of these people, as evidenced by the fact that, although most had lived in the same place since birth, 72% of the adults in the contaminated settlements wanted to relocate. In contrast only 8% in the control settlements wanted to leave.

Toxicology. Blood levels of several toxic materials other than radionuclides were measured. In addition there was particular concern about the possibility of lead poisoning from food and water contaminated by lead that was dumped into the reactor and vaporized by the fire. Levels of cadmium, mercury, and lead were low when compared to those found in the general populations of Italy, Sudan, and the United States.

Hematology. Hematocrit, hemoglobin, erythrocyte number, and erythrocyte mean corpuscular volume were obtained for persons living in control and contaminated settlements. No significant differences were found between control and contaminated settlements. Low hemoglobin levels and low red cell counts were seen in some children. However, no statistically significant difference in values for any age group in either contaminated or control settlements was found. There also was no difference in leukocytes and platelets between the control and contaminated populations nor was there any apparent significant effect on the immune systems of these populations, as determined by the levels of lymphocytes and the prevalence of other diseases.

Thyroid Gland. The thyroid is particularly susceptible to radioiodine and so was of particular interest to the Task 4 investigators. Blood levels of both thyroid stimulating hormone (TSH) and thyroid hormone (free T4) were normal in the children examined. A comparison of these levels from children in the control settlements with those from children in contaminated settlements showed no significant difference. A comparison of thyroid sizes and size distributions of these two groups also showed no differences. The incidence of thyroid nodules in both children and adults was similar to those reported in other countries.

Neoplasms.   The incidence of cancer began rising in Russia before the accident. Task 4 did determine that reporting had been incomplete but could not determine if this was due to methodological differences, improvements in detection, or some other cause. Although the data did not show a striking increase in leukemia or thyroid tumors since the accident, this possibly was the result of several factors, including the classification method used. Thus, an increase in such tumors must be considered a possibility.

Potential Health Effects in the Future.   Overall, there are two future health concerns for the villagers affected by Chernobyl, one more immediate than the other. The first and current effect is the stress produced by the accident itself combined with that induced by the prospect or actuality of relocation. The majority of these people have lived in remote, rural areas for generations, and relocation, even to areas similar in geography and with similar lifestyles, is not easy for them.

The second potential effect is more long-term: the possibility of radiation-induced problems, particularly cancer. How long it will take for this effect to be observed and to what degree it will be seen depend on many factors, such as the duration of exposure, the age at exposure, degree of exposure, means of exposure, and so on. The potential carcinogenic risk is greater for children than it is for adults. Because exposure for the villagers is low and at a low rate, their risk is less for cancer induction than the risk for those people who received high doses in short periods of time (e.g., at the plant). It is nearly impossible to calculate the risk for individuals, because their doses varied so greatly. There will, however, certainly be an increase in the number of cases of leukemia and solid tumors in the next several years. A typical estimate for a village of 10,000 persons is that the number of such deaths is estimated to possibly increase from 1700 to 1750 during the next 10–50 years.

Conclusions

Technological disasters like the accident at the Chernobyl nuclear power plant present unique and enormously complex problems in terms of assessing their effects on health. Unlike natural disasters high-tech accidents often include hazards that are invisible and arcane, and therefore the people affected have different attitudes about the event than they do about more familiar natural disasters. In addition there is a tremendous sense of loss of control. Much of the anxiety seen in the villagers examined by Task 4 physicians, for instance, was the result of no information or conflicting information.

The tremendous increase in adverse health effects reported by much of the media after Chernobyl was not confirmed by the Task 4 investigation. There are two fundamental reasons for this. The first reason is that not enough time has elapsed since the accident for those effects that will become manifest to be seen. The second reason is that the original dose estimates for the people in contaminated villages were much too high. A major factor in the revision of these dose estimates was the ability of the Soviet authorities to bring uncontaminated food to the villages.

The new Confederation of Independent States is undergoing enormous political, social, and economic change. What can and will be done in the future by the government is uncertain. The International Chernobyl Project, in general, and Task 4, in particular, made recommendations concerning the need for more and better equipment, improved methodology, increased education, and continued health care efforts.

Reference

1. Report by an International Advisory Committee. *The International Chernobyl Project: Technical Report: An Assessment of Radiological Consequences and Evaluation of Protective Measures*; International Atomic Energy Association: Vienna, Austria, 1991.

RECEIVED for review August 7, 1992. ACCEPTED revised manuscript April 1, 1993.

Health Studies of U.S. Women Radium Dial Workers

A. F. Stehney

Environmental Research Division, Argonne National Laboratory, Argonne, IL 60439

Follow-up studies of U.S. radium dial workers traced 1322 women first employed between 1913 and 1929, 1403 women first employed between 1930 and 1949, and 744 women first employed between 1950 and 1979. Many women who worked during the earliest period ingested large amounts of radium, because practices such as tipping radium-laden brushes with the lips were not prohibited until about 1925. Early effects included acute anemias and bone destruction. The principal chronic effects among the dial workers were 62 cases of bone sarcoma and 24 cases of carcinoma of the paranasal sinuses or mastoid air cells ("head carcinomas"), but no bone sarcomas or head carcinomas were diagnosed in women who began painting dials after 1925. Deaths caused by leukemias or other specific causes are not established as related to internally deposited radium.

T HE TRAGIC STORY OF THE RADIUM DIAL PAINTERS needs little introduction. The earliest identified manifestation of radium poisoning was a painful and disfiguring destruction of the jaws of some young women who worked at a factory in Orange, New Jersey. Further reports of bone damage, severe anemias, and deaths brought widespread coverage in newspapers of the period, and five of the dial painters sued their former employer in 1927. During subsequent trials news stories made known to a horrified public that radium had become part of the bone structure of these women and that they were probably "doomed to die" early and painful deaths.

0065–2393/95/0243–0169$09.80/0

A similar story of anemias and crippling bone destruction developed somewhat later among employees of a radium dial company in Ottawa, Illinois. During a trial that was settled in 1938, the chilling appellation of "society of the living dead" was coined by a newspaper reporter. The same phrase sometimes appears in the news media when former dial painters die or when residual radioactivity is found at the former site of a radium dial factory or radium refinery.

The purposes of this chapter are to provide a brief review of developments that led to the establishment of a tolerance level for occupational exposure to radium and to summarize quantitative findings about radiation effects that were reported or assumed to occur in the dial workers.

Early Developments

The U.S. luminous radium dial industry began in about 1913 and grew rapidly—from about 8500 dials in 1913 to 2.2 million in 1919 (1). The paint, consisting of zinc sulfide made luminous by the addition of radium and small amounts of impurities, was usually applied to dial hands and numerals by brush. During the early years of the industry, it was common practice for the dial painters to point their brushes by mouth. This practice was probably the principal route by which many of the painters acquired large body burdens of radium, although inhalation of radium-laden dust in the work place is also cited (2, 3).

Indications of possible occupational poisoning among women employed by a radium dial plant in Orange, New Jersey, began to appear in the early 1920s. Severe, nonhealing deterioration of the jawbone of a dial painter was reported by a dentist who, with remarkable prescience, coined the term "radium jaw" in 1924 (4). This report was followed by publication of the results of a medical study that found severe anemias and bone necrosis in several of the women employed at the plant (5). During this period an industrial survey of the plant noted jaw necroses among the dial painters and recommended that the shaping of brush tips by mouth be stopped (3).

The most thorough and longest studies were done by Martland, Medical Examiner of Essex County, New Jersey. Martland et al. (2) were the first to show the presence of radioactivity in the bodies of former dial painters and to blame unequivocally internally deposited radium for the blood dyscrasias and bone necroses that occurred in dial painters. In 1929 Martland and Humphries (6) reported two cases of osteogenic sarcoma among 15 women whose deaths were attributed to radium poisoning; these authors promptly characterized these malignancies as radium induced on the grounds that 2 of 15 "is too large to be passed over as due to coincidence". Martland soon came to the opinion that, in the absence of additional intake of radium, bone sarcomas and radiation osteitis would replace the acute anemias and jaw necroses seen in the earlier cases in New Jersey (7). In effect he distinguished the acute effects of massive doses of radiation from delayed effects and from the chronic effects of continuing irradiation from residual radium in the body. Another major chronic effect, carcinomas of the paranasal sinuses, began to appear somewhat later (8). In 1929 Martland published a remarkably clear and complete account of "occupational poisoning" in the New Jersey radium cases (9), including numbers of persons employed at the plant, the nature of the luminous paint and the application methods, the action of radium on body tissues, findings of deleterious effects, measurements of radioactivity, court cases, newspaper stories and editorials, and community action.

Because of unfavorable publicity, the radium company closed its New Jersey plant in 1926 and moved to New York City, where it continued work on a smaller scale. The following year, after much agitation by community organizations and support from Martland, radium poisoning was declared an occupational disease in New Jersey.

Employees of the New Jersey company were the earliest and most publicized of the radium cases, but there were other companies using radium during this period. Flinn (10) reported severe bone necroses and much radium in the body of a woman who painted dials in Waterbury, Connecticut, and Martland (7) described the case of a woman who died with osteogenic sarcoma 5 years after painting dials in New York and Connecticut plants. A major rival to the New Jersey company operated dial painting plants in Ottawa, Illinois. No serious radium effects among the Illinois painters were reported during the 1920s, and the company claimed that its operations were safe because its luminous paint contained only radium (^{226}Ra) as the activator, whereas the New Jersey company had used a mixture of radium and mesothorium (^{228}Ra) in its paint. However, severe bone necroses in some of the Illinois dial painters became apparent by the end of the decade, and the company discontinued its Illinois operations in the early 1930s. Several former employees of the company sued for compensation in 1938, and the attendant publicity generated much the same fears for the dial painters that had developed in New Jersey 10 years earlier. Interestingly enough another company took over the Illinois dial painting business at a nearby site in 1934 and flourished for many years.

In 1929 the U.S. Department of Labor issued the report of an investigation of 31 plants engaged in radium dial painting or other commercial applications of radioactivity (1) in which it estimated that "not more than 2000 individuals in all" had engaged in luminous dial painting "during the 16 years the industry has existed in the United States". The survey found 23 deaths believed to have been caused by radium poisoning and 19 living persons with effects also attributed to radium. Most of these cases were dial painters from the New Jersey plant. Between June 1929 and March 1930, the U.S. Public Health Service investigated seven factories, which at that time were "all the dial painting factories in the United States" except for a few very small plants (11). Radium painting had gone on for periods varying from 8 to 15 years at the factories, but only 14 men and 228 women still employed at the time of the investigation worked with or were working with radium, and no serious health effects were found in those individuals examined.

World War II brought a considerable upsurge in the luminous dial industry, and by 1941 the U.S. National Bureau of Standards thought it advisable to establish a tolerance value of 0.1 μg of radium fixed in the body in order to protect the "many hundreds of individuals who have entered the dial painting profession during the present war" (12, 13). The tolerance value of 0.1 μg radium (0.1 μCi ^{226}Ra) was based in great part on measurements by Evans and others at the Massachusetts Institute of Technology (MIT) of individuals with 0.5 μg radium or less and no clinical symptoms of radium damage 7–25 years after radium intake. Continuation of these studies resulted in the publication of a definitive paper by Aub et al. in 1952 (14) on the "clinical study of 30 patients . . ." and the "physical aspects of radium and mesothorium toxicity".

Of particular interest to nuclear chemists is the fact that the tolerance level for radium was established at a time when concern about safe standards for internal emitters was about to spread far beyond the radium industry. A 1943 publication by Evans also was very timely because of its many recommendations for safe handling of radioactive materials in the laboratory and for monitoring of radiation exposures of personnel (13).

Methods

Radium Populations and Follow-Up Groups. Comprehensive studies of the health effects of radium in humans were initiated during the 1950s with the support of the U.S. Atomic Energy Commission (AEC). Large numbers of radium-exposed persons were located and studied by the Radioactivity Center (RC) at MIT, by a joint project of Argonne National Laboratory and the Argonne Cancer Research Hospital (ANL–ACRH) and by the New Jersey Radium Research Project (NJRRP) (15) in the New Jersey State Department of Health (1957–1967). The RC was a continuation and expansion of studies by Evans on the measurement and toxicity of radium that dated back to 1934

at MIT. Follow-up efforts of the three projects were somewhat related
to geographic location: the RC tended to concentrate on radium cases
in Connecticut and Massachusetts, ANL–ACRH on cases in Illinois,
and the NJRRP on cases in New Jersey and New York.

At a symposium in 1967 Evans proposed that the radium studies
be combined and continued at a center that would be responsible for
follow-up of all radium-exposed persons and carry out research on radioactivity in humans (*16*). This concept was accepted by the AEC,
and, at the recommendation of an advisory committee, a unified program was established at Argonne National Laboratory as the Center
for Human Radiobiology (CHR). During 1970–1972 the CHR acquired
the case files, or copies, and tissue samples of the three earlier AEC-
supported radium projects. These files provided the names of some
2600 radium-exposed individuals, including not only women dial
painters, but men and women who had been exposed to radium as
laboratory workers, through medical use, and from commercial nostrums. Measurements of radium body burden had been made in 1100
of these persons, and the status of about 700 others had been determined. By June 1984 the CHR and its predecessors had identified
5784 radium-exposed persons by name, determined the radium burdens and health histories of 2374 of these persons, and determined
the vital status of 2401 others among those known by name (*17*). At
the end of 1984 follow-up of radium-exposed persons was sharply reduced, and the CHR was absorbed into the Biological, Environmental, and Medical Research Division of Argonne National Laboratory.

Table I shows numbers of women, by year of first employment,
who were exposed to radium-activated luminous materials in the United
States as dial painters or in other nonlaboratory jobs, such as making
luminous light pulls. Because they had similar types of exposure, the
common practice is to refer to all these women as radium dial workers
or painters unless more precise job distinctions are required and known.

Radium Metabolism and Dosimetry.   Both ^{226}Ra (half-life 1600
yr) and ^{228}Ra (half-life 5.7 yr) are precursors of long radioactive decay
chains that include alpha-particle, beta-particle, and gamma-ray emitters, and each has an isotope of radon in its decay chain (3.82-d ^{222}Rn
from ^{226}Ra and 55.6-s ^{220}Rn from ^{228}Ra). Outside the body the radium
decay chains are hazardous because of gamma-ray irradiation and the
accumulation of radon and radon daughters in the air. Within the body
the radiation doses from beta particles and gamma rays are negligible
compared to that from alpha particles in the decay chains. However,
determinations of the amount of radium in living persons are made
by measurement of gamma rays that escape the body and measurement of radon in the breath (*18*).

**Table I. Numbers and 1985 Status of Women
Radium Dial Workers by Year of First Employment**

Year	Total	Living	Dead	Unknown
1910–1914	19	4	12	3
1915–1919	517	89	303	125
1920–1924	552	159	345	48
1925–1929	471	230	180	61
1930–1934	73	40	32	1
1935–1939	88	63	18	7
1940–1944	1207	654	229	324
1945–1949	477	291	76	110
1950–1954	412	301	64	47
1955–1959	190	163	21	6
1960–1964	104	93	7	4
1965–1969	179	160	6	13
1970–1974	28	26	0	2
1975–1979	1	1	0	0
1910–1979	4318	2274	1293	751

Note: Data from files of radium study, ANL.

Several studies show that about 20% of ingested radium is quickly transferred from the gastrointestinal tract to the blood (*19, 20*), and about the same fraction of inhaled radium is transferred from the lungs to the blood within a few months after intake (*21*). After radium enters the blood a large fraction is excreted in a few weeks, but the rate of excretion diminishes rapidly with time. Whole-body retention of a radium isotope after entry into the blood can be described as a power function of time after entry, as in the empirical relationship obtained by Norris et al. (*22*):

$$R = 0.54t^{-0.52} e^{-\lambda t} \quad (t \geq 1 \text{ day}) \tag{1}$$

Here, λ is the radioactive decay constant of the isotope and R is the fraction remaining in the body at t days after a single injection of radium into the blood. This equation indicates that retention of ^{226}Ra is about 2.5% at 1 year and about 0.33% at 50 years after injection; much less ^{228}Ra would be present at 50 years because of its relatively short physical half-life. From data given by Norris et al. (*22*), one may also estimate that about 60% of the ^{222}Rn produced by ^{226}Ra in the body at 50 years is exhaled. As an alkaline-earth element, radium has metabolic properties similar to calcium and, in particular, deposits preferentially in bone. Numerous analyses of radium in autopsy samples and studies of the metabolism of alkaline earth elements show that the skeleton contains more than 95% of the radium still in the body 20 years after intake (*20, 23*).

The amount of radium swallowed or inhaled by individual dial painters is not known directly, but the amount of radium that enters the blood during the exposure period ("systemic intake") can be estimated from measurements of body burden at later times and extrapolation back to the time of intake by the use of equation 1 or by more detailed retention functions. The number of radioactive transformations accumulated in an organ after intake can then be calculated from the time integral of the organ retention function, and the radiation dose to the organ can be calculated as the product of the number of transformations and the specific effective energy absorbed in the organ per transformation (24). Evans (25) and Evans et al. (26) have described in considerable detail methods of calculating the average skeletal dose ("cumulative rads") from ^{226}Ra and ^{228}Ra.

Dose and dosage parameters that correspond to a residual burden of 1 μCi ^{226}Ra in the body 50 years after exposure to radium are shown in Table II. The values for systemic intake and skeletal dose were calculated with the retention function of Norris et al. (22). However, recent work by Keane et al. (27) indicates that radium is retained less tenaciously at lower levels of systemic intake than at the levels of 70–450 μCi that were studied by Norris et al., so systemic intake and radiation dose relative to a residual burden of 0.1 μCi at 50 years would be greater than the ratios implied in Table II. Some reassessment of radium doses may be necessary, and this possibility is under study at ANL (28). The estimate of soft tissue dose is from a 1983 article by Keane and Schlenker (29).

Computations.   Several of the studies of health effects among women dial workers described in this chapter were concerned with deriving dose–response relationships by fitting equation parameters to observed data. These studies usually employed least-squares fitting routines of the type available in standard statistical packages (30, 31), and statistical significance was determined from the number of degrees of freedom and the sum of the chi-square differences. For these anal-

Table II. Radioactivity and Radiation Doses Associated with the Presence of 1.0 μCi ^{226}Ra (37 kBq) in the Body 50 Years after a Short Period of Intake

Quantity	Classical Units	SI Units
^{222}Rn in breath	10 pCi L^{-1}	370 Bq m^{-1}
Systemic intake (^{226}Ra)	300 μCi	11 MBq
Body intake	1500 μCi	55 MBq
Average skeletal dose (female)	4000 rads	40 Gy
Average soft-tissue dose[a]	10 rads	0.1 Gy

[a]Estimate based on data in reference 29.

yses the figure of merit was the "p value", that is, the probability of a poorer fit from merely statistical fluctuations, and dose–response relationships fitted with p values less than the 0.05 were not considered acceptable.

The other studies of health effects described in this chapter usually compared the observed numbers of cases with the numbers expected from age-, year-, and cause-specific rates for the general population of U.S. white females (32). The results were expressed as standardized mortality ratios (SMR = observed/expected × 100), and statistical significance was shown in various ways. In the summaries of findings given in this chapter, some uniformity of presentation was achieved by showing the observed and expected numbers and using them in tests of significance. The tests were based either on the summary chi-square [$\chi^2 = (|\Sigma \text{ obs} \cdot - \Sigma \text{ exp}.| - 0.5)^2/\Sigma \text{ exp}.$], from which a p value was calculated for 1 degree of freedom (33), or on the Poisson probability (P) that a number equal to or greater than the observed number would have been obtained by chance if the true value was the expected number (34). For these tests small values of p or P were indicative of a significant difference between observed and expected numbers.

Health Effects

Bone Sarcomas and Head Carcinomas. A high incidence of bone sarcomas was noted not only among radium dial painters, but in numerous other populations of persons who accumulated large body burdens of ^{226}Ra or ^{228}Ra (16, 25, 35, 36) or ^{224}Ra (37, 38). A compilation to mid-1988 by Schlenker et al. (39) showed a total of 62 confirmed and three doubtful cases of bone sarcoma among women dial painters in the Argonne radium study (including one confirmed case first diagnosed in 1981 and another in 1988). All of these women entered the dial industry before 1926. The obvious increase of incidence of bone sarcomas with the amount of body radium, the low natural incidence, the absence of life-style factors, the low incidence among atomic-bomb survivors, and the fact that radium deposits preferentially in bone leave little doubt that the high incidence of bone sarcomas in radium-exposed populations was caused by internal deposits of radium.

Characteristics of radium-induced head carcinomas were described in detail by Littman et al. (40). These cancers are thought to be caused by ^{226}Ra but probably not by ^{228}Ra, because they occurred in persons with large burdens of ^{226}Ra and little or no ^{228}Ra but not vice versa (16). The mechanism is believed to involve the accumulation of ^{222}Rn in air spaces adjacent to bony structures in the paranasal sinuses and mastoids (*41*). The 1984 CHR list of "radium-induced malignancies" (*42*) showed a total of 24 women dial painters with head carcinomas, including two cases first diagnosed in the 1980s, all of whom were first employed before 1926.

Because it was well accepted that radium caused bone sarcomas and head carcinomas in radium-exposed persons, the principal scientific objective of studies of these two types of tumors during the last 2 decades was to derive dose–response relationships, with particular attention to the probability of these tumors occurring at very low doses.

Evans (*43*) observed 33 bone sarcomas and 10 head carcinomas in 605 men and women in the combined MIT–NJRRP series for whom radium burdens were known. This group included dial painters, persons who received radium medically or as nostrums, laboratory workers, and others. A total of 102 persons received more than 1000 cumulative skeletal rads, and all tumors of the two types definitely associated with internal radium occurred in these high-dose people. A plot of time from radium exposure to appearance of the 43 radiogenic tumors versus cumulative rads showed a tendency for an increase in time to appearance of tumor as the dose decreased. Evans suggested that this finding supported the concept of a practical threshold dose; that is, at very low doses the time from radiation exposure to appearance of tumor might exceed maximum lifespans. In obtaining a dose–response relationship, Evans combined both types of tumors in a response parameter called "cumulative tumor incidence", that is, the number of tumors of either kind in a dosage cohort divided by the number of people in the cohort. After Evans removed cases he deemed to be "epidemiologically unsuitable" because of possible selection on the basis of symptom, 67 persons with skeletal doses of more than 1000 rads remained, and among them there were 12 cases of bone sarcomas and seven cases of head carcinomas. Dividing these cases into four dose groups in the range of 1000–50,000 rads resulted in a mean value of 28 ± 6% cumulative tumor incidence for doses above 1000 rads and zero incidence below 1000 rads (Figure 1).

Rowland et al. (*44*) examined the incidence of bone sarcomas in radium cases separately from that of head carcinomas and considered only women radium dial workers when deriving quantitative relationships between incidence and radium intake. The study population at the end of 1979 included newly located cases in addition to the cases studied by the earlier radium projects, and radium burdens were measured in 1468 women whose employment in the radium dial industry began before 1950. Among these women were 42 cases of bone sarcoma, all in women first employed before 1930. Rowland et al. tested dose–response relationships by attempting to fit to the radium data

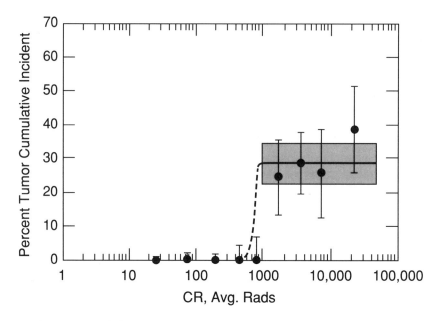

Figure 1. Cumulative occurrence of bone sarcomas and head carcinomas in "epidemiologically suitable" radium cases vs. average skeletal dose. The shaded region corresponds to the mean occurrence of 28 ± 6% between 1,000 and 50,000 rads. (Reproduced with permission from reference 43. Copyright 1974 Health Physics Society.)

equations of the general linear-quadratic-exponential (LQE) form, $I = (C + \alpha D + \beta D^2) e^{-\gamma D}$, where I is the observed incidence, C is the natural incidence, and D is the dose. After the cases were divided into dose groups, equations such as $I = C + \alpha D$, $I = C + \beta D^2$, and so on were fitted to the data points by a weighted least-squares procedure (30, 31), and the goodness of fit was evaluated from the sum of the chi-square differences between calculated and observed values of incidence in the dose groups. The incidence variable was bone sarcoma cases per person-year at risk, and the dose parameter was systemic intake, that is, the calculated quantity (μCi) of radium that entered the blood during the period of exposure (45). Radium intake was used in preference to average skeletal dose, because it allowed the cases to be grouped by intake level (the usual procedure in toxicity studies), it did not vary with time after exposure, and it avoided the possible implication that it represented the (unknown) true dose to critical tissues at risk. For induction of bone sarcomas, ^{228}Ra was weighted by a factor of 2.5 compared to ^{226}Ra; that is, μCi Ra = μCi ^{226}Ra + 2.5 × μCi ^{228}Ra (36). All the bone sarcomas were in women with systemic intake greater than 100 μCi Ra (about 1200 cumulative rads in long-term survivors).*

Rowland et al. (*44*) addressed the question of bias due to symptom-selected cases not by deciding to exclude individual cases, but by defining the population for analysis in two radically different ways. The first method (date of employment) was to accept all women dial painters with measured body burdens and to count years at risk from time of first employment; with these criteria 124 women with a total of 4025 person-years at risk were in the dose groups above 100 μCi that contained the 42 sarcoma cases. The second method (date of first measurement) was to accept only women who had survived at least 2 years without bone sarcoma after first measurement and to count only those person-years occurring after the 2 years; this left 68 women with 890 person-years and 13 bone sarcomas in the dose groups above 100 μCi. The dose group ranges and the total numbers of persons are shown in Table III.

A summary of the results of the least-squares analyses by Rowland et al. (*44*) is given in Table IV. Only dose–response functions that could be fitted to the radium data with positive coefficients and greater than 5% probability of a poorer fit ($p \geq 0.05$ by chi-square test) were accepted. When the test population included all women workers with radium measurements, only the dose-squared exponential form (Figure 2) was acceptable ($p = 0.73$). When the analysis was based on date of first measurement, the dose-squared exponential and the linear form of the dose–response function were both acceptable. Analyses in which the coefficient of the linear term was held constant showed that

Table III. Case Distribution of Women Radium Dial Workers First Employed before 1950 for Whom Radium Measurements Were Made by the End of 1979

Systemic Intake (μCi Ra)[a]	Number of Dose Groups in Analysis	All Women with Measurement		2-Year Survival after Measurement	
		Women	Bone S.[b]	Women	Bone S.[b]
100+	5	124	42	68	13
0.25–99	8	824	0	730	0
<0.25	0	520	0	459	0
Totals	13	1468	42	1257	13

NOTE: Adapted from Table 2 of reference 44.
[a]Systemic intake = μCi ^{226}Ra + 2.5 μCi ^{228}Ra.
[b]Bone sarcomas.

*Two new cases of bone cancer in former radium dial painters were diagnosed after the close of this study at the end of 1979. Average skeletal doses of about 460 and 8100 rads were estimated for the two cases (*39*), but the ^{228}Ra dose to the former possibly was underestimated (*42*).

Table IV. Dose–Response Functions with Acceptable Least-Squares Fits to Incidence of Bone Sarcomas in Women Radium Dial Workers First Employed before 1950 (Positive Coefficients and Chi-Square, $p \geq 0.05$)

Equation	Results
All coefficients free	
Based on year of first employment	$(C = 0.7 \times 10^{-5}\ \mathrm{py}^{-1})$:
$I = (C + \beta D^2)\, e^{-\gamma D}$	$\beta = (7.0 \pm 0.6) \times 10^{-8} \quad p = 0.73$
	$\gamma = (1.1 \pm 0.1) \times 10^{-3}$
Based on date of first measurement	$(C = 1.75 \times 10^{-5}\ \mathrm{py}^{-1})$:
$I = C + \alpha D$	$\alpha = (2.0 \pm 0.6) \times 10^{-5} \quad p = 0.26$
$I = (C + \beta D^2)\, e^{-\gamma D}$	$\beta = (1.8 \pm 0.4) \times 10^{-7} \quad p = 0.27$
	$\gamma = (1.5 \pm 0.2) \times 10^{-3}$

$I = (C + \alpha D + \beta D^2)\, e^{-\gamma D}$ model with only β and γ free
Based on year of first employment:

$\alpha = 0$ (fixed)	to $\quad \alpha = 1.3 \times 10^{-5}$ (fixed)
$\beta = (7.0 \pm 0.6) \times 10^{-8}$	$\beta = (4.3 \pm 1.2) \times 10^{-8}$
$\gamma = (1.1 \pm 0.1) \times 10^{-3}$	$\gamma = (0.9 \pm 0.1) \times 10^{-3}$
$p = 0.58$	$p = 0.05$

SOURCE: Adapted from Table 3 of reference 44.

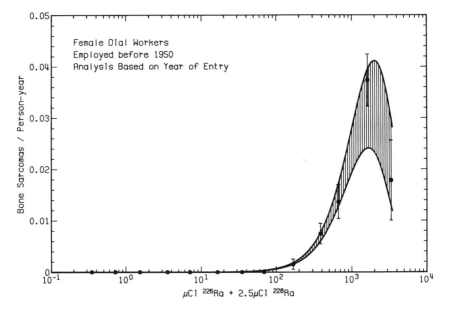

Figure 2. Bone sarcomas per person-year at risk vs. systemic intake of radium in women radium dial workers employed before 1950. The curves show the dose-squared exponential function and the range of values covered by ±1 S.D. of the fitted coefficients. (Reproduced with permission from reference 44. Copyright 1983 Health Physics Society.)

acceptable fits of the complete LQE function to the data for all the cases could be obtained if the coefficient was held to values no more than 1.6×10^{-5} bone sarcomas per person-year per μCi Ra. The coefficient of the linear function fitted to the analysis based on the date of the first measurement had a similar value (2.0×10^{-5}).

In an earlier analysis of dose–response relationships, Rowland and co-workers (36) examined the incidence of head carcinomas among women radium dial painters first employed before 1930 in much the same way as described previously for bone cancers. The study population comprised 749 women whose radium burden was measured. Since ^{228}Ra was known to be much less effective than ^{226}Ra in inducing head carcinomas, the dose parameter (D) for these cancers was defined as systemic intake (μCi) of ^{226}Ra only. There were 17 cases of head carcinoma in 134 women in dose groups above 25 μCi, and no head carcinomas in women with smaller intakes of ^{226}Ra. When the various forms of the dose–response function were tested for fit to head carcinomas per person-year versus μCi ^{226}Ra, only the quadratic form ($I = C + \beta D^2$) and the linear-quadratic form ($I = C + \alpha D + \beta D^2$) were rejected by the least-squares procedure. The simplest form, $I = C + \alpha D$, with $\alpha = (1.6 \pm 0.2) \times 10^{-5}$, provided the best fit. As a means of reducing bias due to nonrandom selection, each case in which the body was exhumed for radioactivity measurement was excluded from the analysis of data. This exclusion resulted in the removal of 23 persons and one head carcinoma from the study, but the results obtained by least-squares were almost identical to those obtained previously. Rowland et al. (36) also showed that, for bone sarcomas and head carcinomas, the same forms of the dose–response function were favored by least-squares tests when the dose parameter was average skeletal dose (rads) as when it was systemic intake (microcurie).

Skeletal Impairment. Destruction of bony tissue was among the earliest deleterious effects ascribed to radium. Blum (4) described mandibular osteomyelitis in a dial painter as radium jaw, and Hoffman (5) published a review of the medical records of dial painters under the title of "Radium (Mesothorium) Necrosis". Martland (9) subsequently determined that bone necroses occurred not only in the jaws, but throughout the body of dial painters and drew attention to clinical similarities of these necroses to what Ewing (46) described as "radiation osteitis" in bones exposed to large doses of radiation from external sources.

Follow-up studies of radium cases usually included radiographic examination of the skeleton for early signs of bone sarcoma or head carcinoma. These X-ray studies revealed changes in bone structure that were characteristic of radium deposition (14, 47), and quantitative measures to rank relative amounts of bone necrosis began being developed in about 1960 (48). Radiologists at ANL–ACRH and the RC then adopted a standardized scoring system to quantify the severity of skeletal damage seen on X-ray plates (25, 49). The types of bone damage scored include coarsening of trabeculation, areas of decreased or increased bone density, spontaneous fractures, and bone or head cancers. Studies done with this scoring system indicated that scores below 5 were of no clinical significance. A so-called "reduced X-ray score" was also defined by considering only changes in trabeculation and bone density. The severity of these changes varies with dose, so the reduced score is a measure of a nonstochastic effect (50).

Keane et al. (49) compared the reduced X-ray scores of two groups of women dial workers: 201 former employees of a plant in Illinois, whose radium intake was predominantly ^{226}Ra (more than six times the intake of ^{228}Ra), and 159 former employees of a plant in Connecticut, whose radium intake was predominantly ^{228}Ra (about seven times more than ^{226}Ra). Figure 3 is a plot of reduced score versus systemic intake

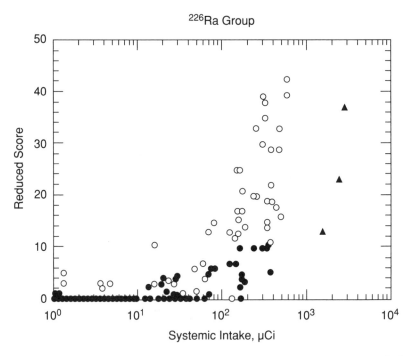

Figure 3. Reduced score versus systemic intake for women radium dial workers with predominantly ^{226}Ra exposure. Reduced score is a measure of non-neoplastic damage to bone. (Reproduced with permission from reference 49. Copyright 1983 International Atomic Energy Agency.)

(μCi ^{226}Ra + μCi ^{228}Ra) for the ^{226}Ra group; the corresponding plot for the ^{228}Ra group was similar in appearance. For both isotopes the scores decreased strongly with decreasing dose to levels deemed "clinically insignificant" (no symptoms expected) at about 25 μCi and to levels found in a group of control subjects at 10 μCi. The spread in reduced score values was greater for ^{226}Ra than for ^{228}Ra, but the mean scores at most intake levels were about the same.

Life Shortening.   Stehney et al. (*51*) examined the survival times of women radium dial workers for evidence of life shortening that could be ascribed to illnesses other than the known radium-related tumors (bone sarcomas and head carcinomas). The study population comprised 1235 women employed in the dial industry before 1930 whose year of birth and vital status at the end of 1976 were known. The analysis used life-table methods to record the time sequence of events in 1-year intervals after the start of employment and compared the observed numbers of deaths with the numbers of deaths expected on the basis of age-time-specific mortality rates for U.S. white women (*32*). The survival probabilities that may be observed in the absence of risk from bone sarcomas and head carcinomas were estimated by methods given by Chiang (*52*), with the assumption of independence among causes of death. Because specific mortality rates were not available for age 85 and older, follow-up time and deaths after age 84 were not counted; this restriction caused 28 women to be withdrawn from the analysis at age 85, including 11 who died at age 85 or later.

A condensed summary of the findings, shown at 10-year intervals in Table V, reveals that deaths from bone sarcomas were three times more numerous than those from head carcinomas and started to occur

Table V. Net Observed and Net Expected Deaths in Pre-1930 Women Radium Dial Workers after Removal of Deaths with Tumor Types Known To Be Radium Related

Interval since Year of Entry (Years)	Number of Persons at Start of Interval[a]	During Interval				
		Deaths with Bone Sarcoma	Deaths with Head Carcinoma	Other Deaths Observed	Other Deaths Expected	Observed Expected
0–9	1235	8	0	48	37.5	1.28
10–19	1179	15	1	45	42.2	1.07
20–29	1118	15	3	44	50.6	0.87
30–39	1055	9	7	72	79.9	0.90
40–49	968	9	4	139	141.5	0.98
50–59	649	2	3	106	106.1	1.00
60–61	37	0	0	1	2.4	0.42
0–61	1235	58	18	455	460.3	0.99

NOTE: Adapted from Table III of reference 51.
[a]Previous number less deaths and live withdrawals during previous interval.

earlier (5 years vs. 19 years). The observed mean and standard error of the survival times of the dial workers from first employment to 59.5 years later was 48.5 ± 0.5 years. Since the expected mean survival during this period was 50.3 years, the mean life shortening in this population was 1.8 ± 0.5 years. Most of the loss of lifetime was due to deaths from bone sarcomas and head carcinomas. Although an excess of other deaths occurred in the first 10 years, the differences between other deaths observed and other deaths expected were not significant at the 5% level in the total or in any of the intervals. However, differences in the early years possibly were reduced by the so-called "healthy worker" effect.

Probabilities of survival in the absence of risk from the known radium-related tumors are plotted at 5-year intervals in Figure 4 as observed and expected cumulative net probabilities of survival. The ratios of observed to expected probabilities (square symbols) indicate less than expected net survival in the early years and more than expected in the later years, but the differences are not statistically significant.

Stehney et al. (51) also examined numbers of deaths in subgroups of dial workers on the basis of radium intake. For this analysis μCi Ra was set equal to μCi ^{228}Ra + μCi ^{228}Ra, and only person-years and deaths that occurred after first measurement of body radium were counted. The results are summarized in Table VI for the 718 dial workers first employed before 1930 whose radium burden was measured while they were living; the average time from first employment to measurement was 40 years, at age 58. Only one death with radium-related tumor occurred at less than 50 μCi, whereas 14 of the 16 deaths in the highest dose group were with bone sarcoma or head carcinoma. The difference between other deaths observed and other deaths expected was not significant at the 5% level for any of the dose groups, but comparison of the 50–500 μCi group with the lowest dose group suggests a dose effect.

Leukemia.   Although it is well known that radiation causes leukemia, that the leukemias begin to appear soon after exposure, and that the red marrow is subject to irradiation from deposits of radium in bone, there is little evidence of excess cases of leukemia among radium dial workers. The autopsy reports and case histories described by Martland and his colleagues in the 1920s and 1930s (6–9) identified no cases of leukemia, although detailed studies of blood and bone marrow were made and severe anemias were observed. Aub et al. (14) reported that one woman in their series of 30 radium cases had leukemia listed as a cause of death. Finkel et al. (35) reported three cases of leukemia, including a case of chronic lymphatic leukemia and the

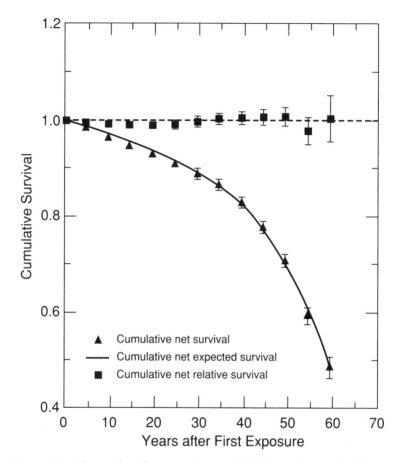

Figure 4. Observed and expected net probabilities of survival of women vs. time (years) after first employment as a radium dial worker, when risk of bone sarcomas and head carcinomas is eliminated. Ratios of observed net survival to expected net survival are shown as square symbols. The error bars represent one standard error. (Reproduced with permission from reference 51. Copyright 1978 International Atomic Energy Agency.)

case mentioned by Aub et al., among 250 women who had painted radium dials in Illinois during the 1920s; the investigators tentatively attributed these leukemia cases to radium deposition, but it is not clear whether this was because of an apparently high incidence or because of an association of bone marrow with radium in bone.

In what was probably the first published study of dial painters that was made with methods commonly employed in epidemiology, Polednak et al. (53) examined the causes of death (from certificates) among

Table VI. Net Observed and Net Expected Deaths in Pre-1930 Women Radium Dial Workers by Radium Dose Group When Entry Is at First Measurement While Living

	Dose Range (μCi Ra Intake)					
	<0.5	0.5–5	5–50	50–500	≥500	Total
Number of persons	202	265	152	80	19	718
Mean date of entry	1965.4	1962.5	1961.0	1957.7	1950.4	1962.2
Mean age at entry	61.1	58.3	57.4	54.3	50.3	58.2
Person-years	2062	3328	2100	1073	144	8707
Deaths	39	63	37	40	16	195
Deaths with						
Bone sarcoma	0	0	0	11	10	21
Head carcinoma	0	0	1	6	5	12
Other deaths observed	39	63	36	24	2	164
Other deaths expected	44.8	64.2	38.8	16.9	2.2	166.9
Observed/Expected	0.87	0.98	0.93	1.42[a]	0.91	0.98

SOURCE: Reproduced with permission from reference 51. Copyright 1978 International Atomic Energy Agency.
[a]Chi-square, $p = 0.20$.

634 women who were identified from employment lists and similar records as having worked in the U.S. radium dial painting industry between 1915 and 1929. Three deaths were coded to leukemia, whereas the number expected from sex-, age-, and year-specific rates was 1.41. In a similar study of women identified from all available sources, Stebbings et al. (54) found three deaths coded to leukemia in 1285 pre-1930 dial workers and four coded to leukemia in 1185 dial workers first employed in 1930–1949, whereas the expected numbers were 4.1 and 1.8, respectively. These findings are not indicative of leukemias caused by internal radium, because the apparent excess of leukemias was observed in the later group, and women in that group generally had much lower radium burdens than did earlier workers.

In a study of leukemia incidence, Spiers et al. (55) included 226 women dial workers first employed in 1950–1969 in addition to those employed before 1950. They found a total of nine incident leukemias in the combined population of 2696 women dial workers, including four cases of chronic lymphatic leukemia, a type not thought to be induced by radiation. The number expected from natural incidence rates was 7.97. Spiers et al. also calculated red marrow doses to 693 women whose radium burdens were measured while they were living and applied risk estimates given by the International Commission on Radiological Protection (56); the numbers of leukemias expected from alpha particles (quality factor of 20), beta particles, and external gamma rays were 2.63, 0.03, and 0.11, respectively. In this population the expected number of natural leukemias was 2.05, so the expected total of natural and radiation-induced leukemias was 4.82—somewhat larger than the actually observed number of two.

Loutit (57) and Mole (58) suggested that cases of aleukemic leukemia possibly were missed in the 1920s. However, Mole also suggested that the continuing absence of excess leukemias is evidence of a low leukemogenic effect from radium in bone.

Breast Cancer. In their study of mortality among 634 women identified from objective records as employed in the radium dial industry before 1930, Polednak et al. (53) found no evidence of excess mortality from breast cancer. About 60% of the women had worked in Connecticut, 30% in Illinois, and 10% in New Jersey. Among all the women there were nine deaths from breast cancer versus 9.45 expected. Among 360 women in the cohort whose radium burdens had been measured, there were three breast cancer deaths (vs. 2.79) in those with less than 50 μCi Ra intake and one death (vs. 0.62) in those with higher systemic intakes of radium.

Adams and Brues (59) studied breast cancer among all women dial workers employed before 1930 whose radium burdens were measured while they were alive (736 women). Follow-up was from year of first measurement to death or 1979. They found excess mortality (6 vs. 1.04 expected) and incidence (9 vs. 2.72) with chi-square p values less than 0.001 in women with radium intakes greater than 50 μCi but no significant difference between observed and expected numbers of breast cancer among women with smaller radium intakes. Analyses by age at first employment, duration of exposure, and parity (no live births vs. one or more live births) showed no relationship between breast cancer and any of these risk factors. From this study Adams and Brues concluded that breast cancer in females may be induced by high levels of internally deposited radium.

Several inconsistencies in findings on breast cancer were noted by Stebbings et al. (54) when they examined cancer mortality in women radium dial workers by place and period of employment in addition to systemic intake of radium and length of employment. Among 1285 women employed before 1930, there were 36 deaths from breast cancer versus 25 expected (chi-square $p < 0.05$) on the basis of U.S. mortality rates and follow-up from year of first employment, but the difference was not significant when adjusted for county rates (30 deaths expected). Extreme variation by work place was found in the mortality from breast cancer of women who worked at the three major places of employment. Many more deaths from breast cancer than expected were reported for women who worked at the Ottawa, Illinois, plant (15 vs. 6.55), more than expected at Orange, New Jersey (13 vs. 7.47), and many fewer than expected at Waterbury, Connecticut (1 vs. 6.67). Review of incidence data (also low at Waterbury) indicated that the quality of ascertainment of breast cancer at Waterbury was as good as at the other locations, so no ready explanation for the deficit at Waterbury is available.

For comparison with internal radium, Stebbings et al. (54) counted breast cancers and person-years in women with measured radium burdens from 1957 or from 2 years after first measurement, whichever was later. Radium-228 was weighted by a factor of 6 relative to ^{226}Ra in soft tissue, because it delivers about six times the radiation dose to soft tissue per μCi of initial intake; however, little change in dose–response comparisons occurred when the two isotopes were weighted equally. Duration of employment was used as a possible indicator of exposure to external gamma-ray radiation and radon and radon daughters in the work place. Because women first employed after 1930 generally had much lower radium burdens than those employed earlier, they were a valuable comparison group for tests of the association of internal radium with induction of cancer.

Numbers of breast cancers in cohorts of women employed before 1930 and first employed in 1930–1949 are shown by place and period of employment in Table VII and by radium intake and duration of employment in Table VIII. The data in both tables are from the article by Stebbings et al. (54). Of special interest is the excess of breast cancers in both cohorts of Illinois women, although they were employed at different factories, and the apparently dose-related excess of breast cancers among women in both time periods, although radium intakes were 10–100 times lower in the 1930–1949 cohort. No relationship to duration of employment was found for either cohort. (Although most of the breast cancers in the later cohort occurred among women who had worked a year or more, most of the women in that cohort had also worked a year or more.) Stebbings et al. (54) also found that, for both cohorts, the excess of breast cancers was not higher in those employed before age 20 and pointed out that this finding was inconsistent with induction by external radiation. Thus, the findings on breast cancer presented somewhat of a puzzle; a significant increase of breast cancers with increase of body radium in the pre-1930

Table VII. Observed and Expected Breast Cancer Deaths in Women Radium Dial Workers, by Factory and First Year of Employment

| | All Cases | | | | Measured Cases (1957+) | | | |
| | <1930 | | 1930–1949 | | <1930 | | 1930–1949 | |
Factory	Obs.	Exp.	Obs.	Exp.	Obs.	Exp.	Obs.	Exp.
Orange, NJ	13	7.48	—	—	1	1.72	—	—
Waterbury, CT	1	6.58	1	0.51	0	3.21	1	0.07
Ottawa 1, IL	15	6.56	0	0.43	8	2.89	0	0.14
Ottawa 2, IL	—	—	9	4.05	—	—	5	1.04

NOTE: Data are from reference 54 and A. F. Stehney, ANL, unpublished data.

Table VIII. Breast Cancer Deaths (1957+) among Measured Women Radium Dial Workers by First Year of Employment, Radium Intake, and Employment Duration

	<1930		1930–1949		<1950[a]	
	Obs.	Exp.	Obs.	Exp.	Obs.	Exp.
Radium intake (μCi)[b]						
<0.5	0	2.02	1	1.11	1	3.13
0.5–4.9	2	2.47	5	0.77	7	3.24
5.0–49	3	2.27	—[c]	—	3	2.27
≥50	6	1.35	—[c]	—	6	1.35
$P(\geq$Obs.$)$[d]	$P(\geq6) = 0.01$		$P(\geq5) = 0.10$		$P(\geq6) = 0.03$	
	(0.005)[e]		(0.046)[e]			
Weeks of employment						
<5	0	1.02	0	0.13	0	1.15
5–49	5	2.90	1	0.37	6	3.27
≥50	6	4.19	5	1.40	11	5.59
$P(\geq$Obs.$)$[d]	$P(\geq6) = 0.54$		$P(\geq5) = 0.50$		$P(\geq11) = 0.25$	

NOTE: Data are from reference 54 and A. F. Stehney, ANL, unpublished data.
[a]This column was added during the present review.
[b]Mesothorium (^{228}Ra) weighted by a factor of 6 relative to ^{226}Ra.
[c]Only four individuals at risk above 5 μCi (0.3 expected total deaths).
[d]Poisson probability of at least the number of observed deaths in the group at highest risk, given the total number observed and the distribution of expected deaths.
[e]Probabilities shown in reference 54.

cohort was seen, but the similar correlation seen at much lower radium levels in the 1930–1949 cohort and other inconsistencies gave reason to doubt that internal radium caused an increase of breast cancers among the dial painters. However, the recalculated probabilities shown in Table VIII and the probabilities calculated for the combined cohorts indicate that the overall distribution of breast tumors is not necessarily inconsistent with an effect of internal radium on breast tumors. Further study is needed to determine whether radiation, lifestyle, environment, or other factors are principally responsible for the apparently elevated number of breast cancers in some groups of radium dial workers and the apparent deficit in the Waterbury workers.

Somewhat ambiguous results also were found in studies of breast cancers among radium luminisers in the United Kingdom, and the findings show the advisability of continued follow-up of dial painters exposed as long ago as World War II. In 1981 Baverstock et al. (60) reported a small excess of breast cancers among U.K. luminisers and suggested that accumulated absorbed doses of more than 20 rads from external gamma rays were the cause, but several years later (1989) Baverstock and Papworth (61) reported that the excess found after continued follow-up was no longer significant at the 5% level. United Kingdom investigators (62) also discounted the possibility that a significant fraction of the breast cancers were caused by the small amounts of internal radium in members of their study population (at levels similar to those found in women who painted dials in the United States during the 1940s).

Rowland et al. (63) compared mortality in U.S. radium dial workers with estimates of external radiation doses. The study population comprised 1261 women first employed in the industry before 1950 whose dates and duration of employment were known. Women who died with bone sarcoma or head carcinoma were excluded. On the basis of reports on levels of gamma-ray radiation and estimates of the amounts of radium in workrooms, Rowland and co-workers assumed that average external radiation dose rates to dial painters were 8 rads/year before 1939 and only 2 rads/year during 1940–1949. Comparison of dose groups of 0–10, 10–50, and 50+ rads showed no trend with dose in the ratio of cancers to deaths or in the distribution of types of cancer. A detailed examination of the amounts and locations of the radium used by the dial painters led to an estimate that the absorbed breast doses were of the order of 15 rads/year. The analysis indicated that breast dose rates at twice the previously estimated external rates might account for half of the ten excess deaths from breast cancer found in the three dose groups combined (26 observed vs. 16 expected).

Multiple Myeloma. Cuzick (64) compiled data from a large number of studies of cohorts exposed to various types of radiation and showed that a nontrivial excess of observed over expected cases of multiple myeloma occurred in most of the cohorts. Included were six deaths of U.S. radium dial painters and seven deaths among Thorotrast patients in Europe. Cuzick's estimates of expected myeloma deaths for the dial painters and Thorotrast patients were 0.86 and 1.86, respectively, indicating that both groups exposed to internal alpha-particle radiation had a significant excess.

Stebbings et al. (54) suggested that Cuzick's estimate of expected numbers of deaths among the dial painters was too low because of a misunderstanding about cause-of-death code numbers and changes in code numbers between the seventh and eighth revisions of the International Classification of Diseases (ICD). However, their analyses still showed a significant excess of multiple myeloma deaths (six observed vs. 2.08 expected, $p < 0.025$) and incidence (seven vs. 2.93, $p < 0.05$) among dial painters first employed before 1950. There were four incident cases, all in the pre-1930 cohort, whose radium burden was measured. Comparison of incidence with radium intake and with duration of employment (Table IX) indicated a closer association with length of employment than with internal radium.

Table IX. Multiple Myeloma Incidence (1957+) among Measured Women Radium Dial Workers First Employed before 1930, by Radium Intake and Employment Duration

Radium Intake (μCi)[a]	Obs.	Exp.	Weeks of Employment	Obs.	Exp.
<5	2	0.47	<50	0	0.35
≥5	2	0.25	≥50	4	0.38
$P(\geq Obs.)$[b]	$P(\geq 2)$	= 0.41	$P(\geq Obs.)$[b]	$P(\geq 4)$	= 0.16
					(0.075)[c]

NOTE: Data are from reference 54.
[a]Mesothorium (^{228}Ra) weighted by a factor of 6.
[b]Poisson probability of at least the number of observed deaths in the group at higher risk, given the total number observed and the distribution of expected incidences.
[c]Probability given in reference 54.

Cancers of Body Sites Directly Exposed to Radium. Like all soft body parts, the lungs and the gastrointestinal (GI) tract are exposed to external radiation and to radiation from the small fraction of body radium in soft tissue. In addition these organs are directly exposed to the radium contained within their walls, the lungs to inhaled radium, radon, and radon daughters (65) and organs of the GI tract to ingested radium and radium being eliminated from the body. The lungs are also exposed to exhaled radon which comes from radium in the body, but this radon is probably only a minor contributor to the lung dose, because it is exhaled too rapidly for accumulation of daughter products. Also, as indicated in Table II, the concentration of ^{222}Rn in the breath of a subject with a large body burden of 1 μCi ^{226}Ra is about 10 pCi L^{-1}, which is not much larger than the average concentration of 3.4 pCi L^{-1} that was reported for a large number of ordinary houses in the United States (66); exhalation of 10 pCi ^{222}Rn L^{-1} would eventually add about 0.3 pCi L^{-1} to the air of a small room.

Polednak et al. (53) found a significant excess of deaths from colon cancer in 634 women radium dial workers first employed before 1930 (ten vs. 4.96 expected, $p < 0.05$ for chi-square), but no excess of deaths from cancers of the stomach, other digestive organs, or the lung. Because most of the radium that enters the body is soon eliminated through the large intestine, the finding of excess colon cancer suggested a causal relationship with radium. However, in a further study of 360 women in the cohort whose radium burden was measured, Polednak et al. found excess colon cancer deaths among 302 women with intake amounts of less than 50 μCi Ra but not among 58 women with higher intakes of radium.

The study of cancers among women dial workers by Stebbings et al. (54) showed that deaths from colon cancer were higher than expected on the basis of U.S. rates (24 observed vs. 15.4 expected, p

< 0.05) in 1285 women employed before 1930, but the difference was not significant when adjusted for county rates (19.2 expected). Deaths from cancers of other digestive organs or from the lung did not differ significantly from the numbers expected. Tests of the distributions of cancers of the stomach, colon, rectum, and lungs versus radium intake and versus duration of employment are summarized in Table X for the 693 women in this cohort with measured radium burdens. The values of P shown in the rows for 50 or more μCi and 50 or more weeks are the Poisson probabilities of the occurrence of the observed number or greater in those rows, given the total number observed and the distribution of expected numbers. No statistically significant dose–response relationships were observed for any of these cancer sites, but there were indications of an effect of radium intake on lung cancer and an effect of duration of employment on cancers of all four sites. Among 1185 dial painters first employed in 1930–1949, a small excess of deaths from colon cancers (eight observed vs. 5.37 expected) and a significant excess of deaths from stomach cancers (seven vs. 1.80 expected) were found.

Cancers of the Central Nervous System.   In a survey of central nervous system (CNS) tumors among radium dial workers, Stebbings and Semkiw (67) found six deaths (vs. 2.84 expected) coded to malignant CNS tumors among 1298 women employed before 1930,

Table X. Observed and Expected Deaths (1957+) in Pre-1930 Measured Women Radium Dial Workers from Cancers of Directly Exposed Sites, by Radium Intake and Employment Duration

	Stomach		Colon		Rectum		Lung	
	Obs.	Exp.	Obs.	Exp.	Obs.	Exp.	Obs.	Exp.
Radium intake (μCi)[a]								
<0.5	0	0.48	3	1.65	0	0.37	0	1.02
0.5–4.9	2	0.72	3	2.45	2	0.55	0	1.52
5.0–49	2	0.43	3	1.50	1	0.34	4	0.94
≥50	1	0.20	0	0.66	0	0.15	2	0.40
$P(\geq\text{Obs.})$[b]	$P(\geq 1) = 0.44$		$P(\geq 0) = 1.0$		$P(\geq 0) = 1.0$		$P(\geq 2) = 0.12$	
Weeks of employment								
<5	0	0.23	0	0.78	0	0.17	0	0.51
5–49	1	0.63	3	2.18	0	0.49	0	1.40
≥50	4	0.98	6	3.30	3	0.74	6	1.96
$P(\geq\text{Obs.})$[b]	$P(\geq 4) = 0.23$		$P(\geq 6) = 0.31$		$P(\geq 3) = 0.15$		$P(\geq 6) = 0.09$ (0.017)[c]	

NOTE: Data are from reference 54 and A. F. Stehney, ANL, unpublished data.
[a]Equal weights for ^{226}Ra and ^{228}Ra.
[b]Poisson probability of at least the number of observed deaths in the group at highest risk, given the total number observed and the distribution of expected deaths.
[c]Probability given in reference 54.

three deaths (vs. 1.79) among 1007 women first employed in 1930–1944, and one death (vs. 1.22) among 1207 women first employed after 1944. However, study of the medical records showed that one of the deaths in the pre-1930 cohort actually involved invasion of a mastoid carcinoma into the brain, and no mention of CNS tumor was found in the hospital records and autopsy report of another case in the same cohort. Three (vs. 0.77 expected) of the four authenticated cases in the pre-1930 cohort occurred among 331 women who worked in Ottawa, Illinois, and two (vs. 0.54) of the 1930–1944 cases occurred among 335 women who worked at a different plant in the same town.

Stebbings and Semkiw also found no malignant CNS tumors reported among 190 women exposed to radium as laboratory workers or as patients who had received radium for supposed therapeutic effects. On the basis of these findings and literature reports of low sensitivity of the nervous system to radiation, these investigators speculated that factors other than radiation might be involved in the etiology of the CNS tumors, but they also pointed out that an unusually high proportion of deaths from CNS tumors in the pre-1930 cohort had occurred outside the brain near bone (four out of a total of seven deaths authentically coded to all types of CNS tumors)—a finding suggestive of an effect from skeletal deposits of radium.

Summary and Discussion

Early studies indicated that bone sarcomas, head carcinomas, and destructive bone changes were likely to be the principal chronic effects of internally deposited radium, and follow-up studies of about 2500 women who worked in the U.S. radium dial industry before 1950 confirmed and quantified these findings. New cases of bone sarcoma and head carcinoma continued to appear among the dial workers into the 1980s, and by 1989 there were 62 known cases of bone sarcomas and 24 cases of head carcinomas, all of them in women who had started work as dial painters before 1926. The incidence of bone sarcomas versus systemic intake of radium was best fitted as a dose-squared relationship, but a low-dose linear component of about 1.3×10^{-5} sarcomas per person-year per microcurie of Ra intake could be added. The head carcinomas were best fitted by a linear relationship of $(1.6 \pm 0.2) \times 10^{-5}$ py^{-1} μCi^{-1}.

Rowland et al. (*44*) used the dose–response equations fitted to the bone sarcomas to estimate the risk from 1 year's consumption of water that contained 5 pCi ^{226}Ra L^{-1}, the maximum level proposed by the U.S. Environmental Protection Agency. Based on 2.2 L/day and 21% absorption of radium from the gut, the systemic intake in 1 year would be 843 pCi and the corresponding lifetime risk would be 1×10^{-8}

bone sarcoma/person-year if the low-dose linear component were used and only 5×10^{-14} py^{-14} if the pure dose-squared relationship were used. The risk of head carcinomas would be about 1.3×10^{-8} py^{-1}. These calculations illustrate the obvious: there may not be a dose ("threshold") below which effects do not occur, but there are doses below which the risk of an effect is negligible.

From the study of radiographs, clinically significant bone changes were found at systemic intakes of more than about 25 μCi ^{226}Ra or ^{228}Ra. Intakes at these levels correspond to the smallest amounts of radium proven to cause deleterious effects in humans, so comparison with the occupational tolerance value of 0.1 μCi radium in the body (12) is in order. Rowland and Lucas, (68) used the retention equation developed by Marshall et al. (20) to show that maximum systemic intake of ^{226}Ra over a 50-year period without exceeding 0.1 μCi at any time during this period was 16.6 μCi. Thus, the radium standard still retains credibility 50 years after its formulation, if one keeps in mind that the requirement is never to exceed 0.1 μCi. It is of interest to note that the apparently independently derived limit on annual intake (1.9 μCi ^{226}Ra) proposed by the International Commission on Radiological Protection (69) implies a total systemic intake of 19 μCi and maximum body burden of 0.12 μCi in 50 years (20, 68).

Examination of the survival times of 1235 women first employed in the radium dial industry before 1930 showed average life shortening of 1.8 ± 0.5 years from date of employment to 59.5 years later. Nearly all of the loss could be ascribed to early deaths from bone sarcomas and head carcinomas, and the life spans of dial workers who did not suffer these malignancies were about the same as those of women contemporaries in the general U.S. population. These findings indicate that, contrary to public perception, most dial painters lived normal life spans, even those who had worked during the period of highest exposures to radium.

Detailed studies showed that leukemia incidence and mortality in the radium dial workers were not greater than expected on the basis of year- and age-specific rates for U.S. white women. Elevated levels of multiple myeloma and breast cancer were found, but examination of related factors led to doubts of a causal relationship with internal radium. Multiple myelomas were better correlated with duration of employment (a surrogate for exposure to external radiation and radon) than to systemic intake of radium, but no relationship to duration was found for breast cancers. Further study of multiple myelomas and breast cancers in dial workers is needed.

Cancers of body sites directly exposed to radium (lungs, stomach, colon, and rectum) were not significantly correlated with either radium intake or employment duration, but stomach, colon, and rectal cancers increased with duration and lung cancers increased with radium intake and with duration.

Findings on CNS tumors in radium-dial workers were not indicative of a radium or radiation effect, but an unusually high proportion was noted of CNS tumors that occurred near bone outside the brain.

In conclusion it may be said that numerous follow-up studies failed to prove any radiation effects on radium dial workers other than bone damage and the bone-related malignancies that first focused attention on radium hazards more than 60 years ago. These effects obviously are the direct result of irradiation from internal deposits of radium in the skeleton. However, other parts of the body receive some radiation from internal radium and its daughter products, and some dial workers were probably exposed to nontrivial amounts of external radiation and radon in the work place. Therefore, other health effects possibly were present but at incidences too small to prove in this limited population. As mentioned earlier, further study of breast cancers and multiple myelomas in the dial workers may be worthwhile. Somewhat paradoxically, leukemia, almost the very embodiment of a radiation effect, probably is the effect most convincingly established as absent among the dial workers. Among Japanese atomic-bomb survivors, leukemias are the most prominent radiation effect and bone cancers probably the least prominent (70).

Acknowledgments

T. J. Kotek provided information on the numbers and current status of women dial workers from data files of the radium study at Argonne National Laboratory, and L. L. Westfall assisted with tables and figures. I am also grateful to A. T. Keane, R. E. Rowland, and J. Rundo for reviewing the manuscript and to K. L. Haugen for valuable editorial assistance.

This work was supported by the Assistant Secretary for Environment, Safety, and Health, Office of Epidemiology and Health Surveillance, U.S. Department of Energy, under Contract No. W-31-109-ENG-38.

References

1. U.S. Bureau of Labor Statistics; *Monthly Labor Rev.* **1929**, *28*, 1208.
2. Martland, H. S.; Conlon, P.; Knef, J. P. *J. Am. Med. Assoc.* **1925**, *85*, 1769.
3. Castle, W. B.; Drinker, K. R.; Drinker, C. K. *J. Ind. Hyg.* **1925**, 7, 317.
4. Blum, T. *J. Am. Dent. Assn.* **1924**, *11*, 802.
5. Hoffman, F. L. *J. Am. Med. Assn.* **1925**, 85, 961.
6. Martland, H. S.; Humphries, R. E. *Arch. Path.* **1929**, 7, 406.
7. Martland, H. S. *Am. J. Cancer* **1931**, 15, 2435.

8. Martland, H. S. *Occupational Tumours. (6) Bones. Encyclopedia of Health and Hygiene*; International Labor Office: Geneva, Switzerland, 1939; pp 1–23.
9. Martland, H. S. *J. Am. Med. Assoc.* **1929**, *92*, 466.
10. Flinn, F. B. *Laryngoscope* **1927**, *37*, 341.
11. Schwartz, L.; Knowles, F. L.; Britten, R. H.; Thompson, L. R. *J. Ind. Hyg.* **1933**, *15*, 362.
12. National Bureau of Standards. *Handbook H27*; U.S. Department of Commerce: Washington, DC, 1941.
13. Evans, R. D. *J. Ind. Hyg. and Toxicol.* **1943**, *25*, 253.
14. Aub, J. C.; Evans, R. D.; Hempelmann, L. H.; Martland, H. S. *Medicine* **1952**, *31*, 221.
15. Sharpe, W. D. *Environ. Res.* **1974**, *8*, 243.
16. Evans, R. D. In *Delayed Effects of Bone-Seeking Radionuclides*; Mays, C. W.; Jee, W. S. S.; Lloyd, R. D.; Stover, B. J.; Dougherty, J. H.; Taylor, G. N., Eds.; University of Utah: Salt Lake City, UT, 1969; pp 191–194.
17. Rundo, J.; Keane, A. T.; Lucas, H. F.; Schlenker, R. A.; Stebbings, J. H.; Stehney, A. F. In *The Radiobiology of Radium and Thorotrast*; Gössner, W.; Gerber, G. B.; Hagen, U.; Luz, A., Eds.; Supplements to Strahlentherapie; Urban & Schwarzenberg: Munich, 1986; Vol. 80, pp 14–21.
18. Toohey, R. E.; Keane, A. T.; Rundo, J. *Health Phys.* **1983**, *44* (Suppl. 1), 323.
19. Maletskos, C. J.; Keane, A. T.; Telles, N. C.; Evans, R. D. In *Delayed Effects of Bone-Seeking Radionuclides*; Mays, C. W., Jee, W. S. S.; Lloyd, R. D.; Stover, B. J.; Dougherty, J. H.; Taylor, G. N., Eds.; University of Utah: Salt Lake City, UT, 1969; pp 29–49.
20. Marshall, J. H.; Lloyd, E. L.; Rundo, J.; Liniecki, J.; Marotti, G.; Mays, C. W.; Sissons, H. A.; Snyder, W. S. *Health Phys.* **1972**, *24*, 125.
21. International Commission on Radiological Protection; *ICRP Publication 30, Part 1*; Pergamon: Oxford, United Kingdom, 1979; pp 23–29.
22. Norris, W. P.; Speckman, T. W.; Gustafson, P. F. *Am. J. Roentgenol. Radium Ther. Nucl. Med.* **1955**, *73*, 785.
23. Schlenker, R. A.; Keane, A. T.; Holtzman, R. B. *Health Phys.* **1982**, *42*, 671.
24. International Commission on Radiological Protection; *ICRP Publication 30, Part 1*; Pergamon: Oxford, United Kingdom, 1979; pp 12–19.
25. Evans, R. D. *Br. J. Radiol.* **1966**, *39*, 881.
26. Evans, R. D.; Keane, A. T.; Kolenkow, R. J.; Neal, W. R.; Shanahan, M. M. In *Delayed Effects of Bone-Seeking Radionuclides*; Mays, C. W.; Jee, W. S. S.; Lloyd, R. D.; Stover, B. J.; Dougherty, J. H.; Taylor, G. N., Eds.; University of Utah: Salt Lake City, UT, 1969; pp 157–191.
27. Keane, A. T.; Rundo, J.; Essling, M. A. *Health Phys.* **1988**, *54*, 517.
28. Thomas, R. G. Argonne National Laboratory; personal communication, 1991.
29. Keane, A. T.; Schlenker, R. A. *Health Phys.* **1983**, *44* (Suppl. 1), 81.
30. Barr, A. J.; Goodnight, J. H.; Sall, J. P.; Helwig, J. T. *A User's Guide to SAS 79*; SAS Institute: Raleigh, NC, 1979.
31. Gallant, A. R. *Am. Stat.* **1975**, *29*, 73.
32. Monson, R. R. *Comput. Biomed. Res.* **1974**, *7*, 325.
33. Mantel, N.; Haenszel, W. *J. Natl. Cancer Inst.* **1959**, *22*, 719.

34. Ginevan, M. E. *Environ. Res.* **1981,** *25,* 147.
35. Finkel, A. J.; Miller, C. E.; Hasterlik, R. J. In *Delayed Effects of Bone-Seeking Radionuclides*; Mays, C. W.; Jee, W. S. S.; Lloyd, R. D.; Stover, B. J.; Dougherty, J. H.; Taylor, G. N., Eds.; University of Utah: Salt Lake City, UT, 1969; pp 195–225.
36. Rowland, R. E.; Stehney, A. F.; Lucas, H. F., Jr. *Radiat. Res.* **1978,** *76,* 368.
37. Spiess, H. In *Delayed Effects of Bone-Seeking Radionuclides*; Mays, C. W.; Jee, W. S. S.; Lloyd, R. D.; Stover, B. J.; Dougherty, J. H.; Taylor, G. N., Eds.; University of Utah: Salt Lake City, UT, 1969; pp 227–247.
38. Mays, C. W.; Spiess, H. In *Radiation Carcinogenesis: Epidemiology and Biological Significance*; Boice, J. D., Jr.; Fraumeni, J. F., Jr., Eds.; Raven: New York, 1984; pp 241–252.
39. Schlenker, R. A.; Keane, A. T.; Unni, K. K. In *Risks from Radium and Thorotrast*; Taylor, D. M.; Mays, C. W.; Gerber, G. B.; Thomas, R. G., Eds.; British Institute of Radiology: London, 1989; pp 55–65.
40. Littman, M. S.; Kirsh, I. E.; Keane, A. T. *Am. J. Roentgenol.* **1978,** *131,* 773.
41. Schlenker, R. A. *Health Phys.* **1983,** *44,* 556.
42. Center for Human Radiobiology; *Environmental Research Division Annual Report*; ANL-84-103 Part II; Argonne National Laboratory: Argonne, IL, 1985; pp 181–185.
43. Evans, R. D. *Health Phys.* **1974,** *27,* 497.
44. Rowland, R. E.; Stehney, A. F.; Lucas, H. F. *Health Phys.* **1983,** *44* (Suppl. 1), 15.
45. Stehney, A. F. *Radiological Physics Division Annual Report*; ANL-7860 Part II; Argonne National Laboratory: Argonne, IL, 1971; pp 9–15.
46. Ewing, J. *Acta Radiol.* **1926,** *6,* 399.
47. Looney, W. B.; Hasterlik, R. J.; Brues, A. M.; Skirmont, E. *Am. J. Roentgenol. Radium Ther. Nucl. Med.* **1955,** *73,* 1006.
48. Finkel, A. J.; Miller, C. E.; Hasterlik, R. J. In *Radiation-Induced Cancer*; International Atomic Energy Agency: Vienna, Austria, 1969; pp 183–202.
49. Keane, A. T.; Kirsh, I. E.; Lucas, H. F.; Schlenker, R. A.; Stehney, A. F. In *Biological Effects of Low-Level Radiation*; International Atomic Energy Agency: Vienna, Austria, 1983; pp 329–350.
50. International Commission on Radiological Protection; *ICRP Publication 26*; Pergamon: Oxford, United Kingdom, 1977; pp 2–5.
51. Stehney, A. F.; Lucas, H. F., Jr.; Rowland, R. E. In *Late Biological Effects of Ionizing Radiation*; International Atomic Energy Agency: Vienna, Austria, 1978; pp 333–351.
52. Chiang, C. L. *Introduction to Stochastic Processes in Biostatistics*; Wiley: New York, 1968.
53. Polednak, A. P.; Stehney, A. F.; Rowland, R. E. *Am. J. Epid.* **1978,** *107,* 179.
54. Stebbings, J. H.; Lucas, H. F.; Stehney, A. F. *Am. J. Ind. Med.* **1984,** *5,* 435.
55. Spiers, F. W.; Lucas, H. F.; Rundo, J.; Anast, G. A. *Health Phys.* **1983,** *44* (Suppl. 1), 65.
56. International Commission on Radiological Protection. *ICRP Publication 26*; Pergamon: Oxford, United Kingdom, 1977; p 10.
57. Loutit, J. F. *Br. J. Cancer* **1970,** *24,* 195.

58. Mole, R. H. In *Risks from Radium and Thorotrast*; Taylor, D. M.; Mays, C. W.; Gerber, G. B.; Thomas, R. G., Eds.; British Institute of Radiology: London, 1989; pp 153–158.
59. Adams, E. E.; Brues, A. M. *J. Occup. Med.* **1980**, *22*, 583.
60. Baverstock, K. F.; Papworth, G. D.; Vennart, J. *Lancet* **1981**, *i*, 430.
61. Baverstock, K. F.; Papworth, G. D. In *Risks from Radium and Thorotrast*; Taylor, D. M.; Mays, C. W.; Gerber, G. B.; Thomas, R. G., Eds.; British Institute of Radiology: London, 1989; pp 72–76.
62. Baverstock, K. F.; Vennart, J. *Health Phys.* **1983**, *44* (Suppl. 1), 15.
63. Rowland, R. E.; Lucas, H. F.; Schlenker, R. A. In *Risks from Radium and Thorotrast*; Taylor, D. M.; Mays, C. W.; Gerber, G. B.; Thomas, R. G., Eds.; British Institute of Radiology: London, 1989; pp 67–72.
64. Cuzick, J. *N. Engl. J. Med.* **1981**, *304*, 204.
65. Bloomfield, J. J.; Knowles, F. L. *J. Ind. Hyg.* **1933**, *15*, 368.
66. White, S. B.; Bergsten, J. W.; Alexander, B. V.; Rodman, N. F.; Phillips, J. L. *Health Phys.* **1992**, *62*, 41.
67. Stebbings, J. H.; Semkiw, W. In *Risks from Radium and Thorotrast*; Taylor, D. M.; Mays, C. W.; Gerber, G. B.; Thomas, R. G., Eds.; British Institute of Radiology: London, 1989; pp 63–72.
68. Rowland, R. E.; Lucas, H. F., Jr. In *Radiation Carcinogenesis: Epidemiology and Biological Significance*; Boice, J. D., Jr.; Fraumeni, J. F., Jr., Eds.; Raven: New York, 1984; pp 231–240.
69. "Limits for Intakes of Radionuclides by Workers"; International Commission on Radiological Protection; *ICRP Publication 30, Part 1*; Pergamon: Oxford, United Kingdom, 1979; pp 98–99.
70. National Research Council, Committee on the Biological Effects of Ionizing Radiations (BEIR V); National Academy Press: Washington DC, 1990; pp 242 and 307.

RECEIVED for review August 7, 1992. ACCEPTED revised manuscript March 25, 1993.

Environmental and Occupational Exposure

Evaluating Health Risks in Communities near Nuclear Facilities

A. James Ruttenber

Department of Preventive Medicine and Biometrics, University of Colorado School of Medicine, Denver, CO 80217

Recent critical reviews of published epidemiologic studies suggest that these studies are not capable of evaluating causal relations between exposure and disease in communities near nuclear facilities. In the United States the combination of dose reconstruction and risk assessment is being tried as an alternate method for assessing health risks. This chapter reviews the limitations of epidemiologic studies, outlines the process of dose reconstruction and risk assessment, and describes a number of dose reconstruction projects underway in the United States.

NUCLEAR FACILITIES AROUND THE WORLD are subjected to intense public scrutiny regarding health risks from past, current, and future operations. Over the years the most popular response to public concern by public health researchers was to conduct epidemiologic studies in populations near these facilities. This approach resulted in numerous studies of disease rates in these communities (Figure 1). Taken separately or as a group, these epidemiologic studies do not provide much clarity regarding the relation between nuclear plants and health risks.

Dose reconstructions and risk assessments are being conducted by independent scientists for communities near nuclear facilities in the United States in an attempt to estimate health risks more accurately

0065–2393/95/0243–0201$08.00/0

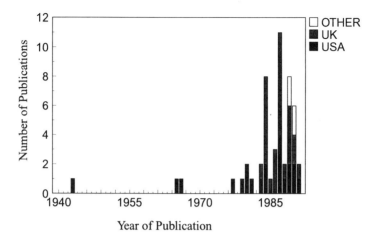

Figure 1. Cancer studies in populations near nuclear facilities.

than with epidemiologic studies. Similarly, such techniques are used to evaluate health risks from the Chernobyl reactor explosion in the former Soviet Republics and in Eastern European countries.

This chapter summarizes the weaknesses of epidemiologic studies of populations around nuclear facilities and describes how modern techniques of dose reconstruction and risk assessment can be used to help clarify health risks.

How Useful Are Epidemiologic Studies?

A recent review of epidemiologic studies around nuclear facilities indicates that most studies were not able to establish or reject causal relations between exposure and disease in nearby communities (1). Commonly, these studies failed to establish dose–response relations between environmental exposure and disease—perhaps the most important of the criteria used to demonstrate causal relations (2). The major reason the dose–response analyses were inadequate was the absence of reliable estimates of exposure or dose.

Most epidemiologic studies estimate exposure with a combination of physical distance and geopolitical boundaries. A common practice is to draw a series of concentric circles around a nuclear facility and base exposure on whether geopolitical units, such as counties or administrative districts, are within these circles. Another popular approach is to use the linear distance from a nuclear facility to the address of a research subject as an indicator of exposure status.

These measures are inadequate for two reasons: (1) the dynamics of ecosystems do not follow political boundaries, and distance is only one of the many factors that contribute to exposure; and (2) the location of the residence of a research subject at the time a disease is diagnosed may not be the same as the residence location for the period when the disease was induced by an environmental exposure.

Studies of the dispersal of radionuclides and the distribution of radiation doses from the Three Mile Island (TMI) nuclear power plant accident (3, 4), the early operations of the Hanford nuclear facility (5–7), and the Chernobyl nuclear power plant accident (8) showed that the distribution of radiation doses, as estimated from models and environmental measurements, did not conform to concentric circles. In fact researchers studying the health impacts of the TMI accident proposed that distance from a nuclear facility can be used as a measure of stress in the population following a nuclear accident, not exposure to ionizing radiation (9).

In a study of cancer in counties near nuclear facilities in the United States (10), four of the nine counties selected as the unexposed controls for the Hanford nuclear facility in southeastern Washington actually received exposure from Hanford via both atmospheric and food chain pathways (5–7). Such an error in classification usually makes false-negative results more likely than false-positive ones.

Another problem common to epidemiologic studies is low statistical power—a measure of the probability that a finding of no association between exposure and disease is actually correct. Statistical power increases with the size of the population in a study, the size of disease risk associated with the magnitude and duration of exposure, and the prevalence of the disease of interest in the unexposed population. Because nuclear facilities are usually located in rural areas and because offsite exposures are generally low, most epidemiologic studies have predictably low power. It is important, therefore, to report the statistical power of such studies if no association is found between exposure and disease—a practice that is rare in the scientific literature. The results from studies with low power or from studies with power that is not reported should not be interpreted as evidence for no effect between exposure and disease.

An example of this problem is the aforementioned study of cancer mortality in counties near nuclear facilities in the United States (10). This study found no suggestion of a risk in counties near nuclear facilities and concluded that, if any excess cancer risk was present, it was too small to be detected with the methods employed. Although the authors acknowledged that they did not prove the absence of any effect, they did not report the power of their study so that readers could estimate the likelihood of accepting a null hypothesis when it should have been rejected.

The reported exposures around nuclear facilities during normal operating conditions usually produce low risks for cancer in surrounding communities, and these populations are usually too small to reliably assess disease, even with the best of epidemiologic techniques. These limitations can be predicted before beginning a study and should not be used as excuses for not being able to interpret results.

Using epidemiologic studies to screen for health risks and to guide future research is a flawed strategy. Because of the limited statistical power of epidemiologic studies, negative findings cannot reliably rule out the presence of health risks. Moreover, multiple studies with flawed or limited methods are no more convincing than a single study with a good design—particularly in the field of radiation biology where such studies must be compared with a number of well-designed ones. The criteria scientists use for establishing causal relations (2, 11) do not include recognizing quantity as superior to quality.

The fact that the scientific community has not accepted the results from existing community studies supports this assessment. In spite of the many studies identifying disease increases in populations near nuclear facilities (1), the health risks are still being debated. Moreover, the evidence from community epidemiologic studies supporting the risks from nuclear facilities is not convincing enough to be used by the national and international agencies responsible for evaluating and setting radiation protection standards.

On the other hand, multiple analyses showing no risk do not convince the public or scientists of the absence of any risk. In the case of the study of all nuclear facilities in the United States (10), critics can easily agree with the authors' own stated limitations. It is interesting to note that the one positive finding in this study—a slight but significant elevation in the incidence of leukemia around the Millstone nuclear power plant in New London County, Connecticut—has spurred additional epidemiologic analysis in communities around this facility, while the nuclear facility with the most convincing evidence for off-site exposure and health risk—the Hanford facility during its early operational years (5-7, 12)—showed no epidemiologic evidence of increased health risks in surrounding communities (10).

The primary reason epidemiologic studies are difficult to interpret is because they usually set out to evaluate simultaneously two alternatives to the null hypothesis: (1) environmental exposures are higher than reported, and (2) the risks from radiation are higher than currently acknowledged. For studies that have not explicitly measured exposure, it is impossible to determine which of the two hypotheses is supported by evidence of increased disease in communities around nuclear facilities.

In spite of the limitations of epidemiologic studies of populations around nuclear facilities, scientists and the public still advocate their use. Some argue that even more studies of these communities are needed to help clear up the debate about whether nuclear facilities present health risks to the public:

> After an initial report of a cluster of childhood leukemia near one nuclear plant in northern England, subsequent investigations revealed that there is a consistent pattern in the United Kingdom of a small but elevated risk of leukemia for children living near nuclear establishments. These clusters differ from many others investigated in that a source of environmental contamination exists near the initial cluster, and it is possible to identify other similar sources of contamination elsewhere for further hypothesis testing.
>
> Identifying several similar sources of environmental contamination and examining for consistent increases in disease rates around them may, in general, be more likely to yield interpretable epidemiologic results than in-depth studies of isolated clusters. In doing this, however, clear prior hypotheses need to be specified before the investigation begins (*13*).

In summary, advocates for epidemiologic studies take two positions: (1) the studies, no matter how limited by methodology, help determine whether additional research is needed; and (2) although a single study may not produce convincing results, many studies around different nuclear facilities, analyzed separately or combined, will improve our knowledge of health risks from these facilities.

Dose Reconstruction and Risk Assessment: Useful Alternatives to Epidemiologic Studies

Dose reconstruction provides estimates of health risk based on estimates of radiation exposure that depend, in part, on the distribution of radionuclides in the environment. Quantitative information about radiation exposure from nuclear facilities to persons living around them helps solve the three biggest problems of epidemiologic studies: dose–response analysis, misclassification bias, and low statistical power.

The steps in dose reconstruction are outlined in Figure 2. The type of dose reconstruction described in this chapter involves state-

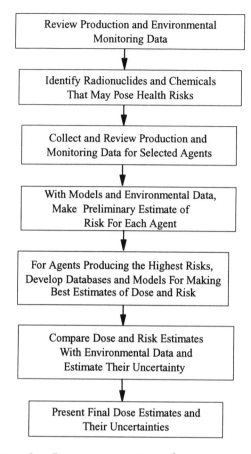

Figure 2. Dose reconstruction: the major steps.

of-the-art techniques applied to site-specific conditions and is far more detailed than techniques of dose estimation used for many risk assessments, such as those for compliance with the Resource Conservation and Recovery Act (RCRA) and the Comprehensive Environmental Response, Compensation, and Liability Act (CERCLA) (14). I will use the terms *dose estimation* and *dose evaluation* to apply to these less detailed approaches and the term *dose reconstruction* for more detailed analyses.

The first step in a dose reconstruction is a thorough review of production and environmental monitoring data to identify radionuclides and chemicals that may pose health risks. Environmental monitoring data, appropriately collected and analyzed, are the best for reconstructing doses. Because such data usually are not available, particularly for early years of operation, dose reconstruction must rely heavily on models of radionuclide movement in the environment.

These models usually depend on a source term—the quantity of radioactive material released to the environment over a defined period of time. Source terms are estimated by extensively reviewing data from production records, recordings from stack monitoring, and mass-balance analysis, a technique that compares quantities of starting materials with the quantities of finished products.

The radionuclides and chemicals that may provide off-site exposures are then assessed with screening models that assume conditions that maximize doses and risk estimates to the public in order to select the ones that require more detailed modeling and uncertainty analysis.

The screening process can involve different levels of effort, depending on the particular site and on the availability of funds. One popular approach is to establish a level of risk below which there is little concern. Doses from chemicals and radionuclides that result in risks above this level receive more careful analysis, while those producing lower risks receive less attention.

The models used in dose reconstruction incorporate the latest advances in modeling techniques and rely on large data sets obtained by in-depth investigations. The increased availability of large microcomputers and advanced software has stimulated improvement in, and access to, state-of-the-art models. These models are designed to describe atmospheric dispersion of agents based on local and regional weather patterns, terrain, and the release characteristics of the facility. Models are also available for the surface, groundwater, and food chain transport of radionuclides and chemicals.

Dose reconstructions can be performed for toxic chemicals as well as for radionuclides, although the risk for cancer per unit dose is usually computed differently for chemicals than for radionuclides. Furthermore, there is little guidance in the scientific literature for determining the combined risk from chemicals and radionuclides.

In order to be meaningful from a public health standpoint, the doses estimated in the reconstruction process are converted to estimates of disease risk. This conversion is accomplished by multiplying the doses by estimates of disease risk per unit dose, which are available in the scientific literature for cancer, genetic damage to offspring from radiation exposure, and some acute and chronic diseases caused by chemical exposures. When using dose reconstruction results for assessing health risks, extensive analyses of the combined uncertainty from all variables are required to establish the upper and lower bounds for risk as well as to estimate the median risk for the exposed group (15).

Independent Scientific Oversight and Public Involvement

In the past, nuclear facilities usually did not provide the public with verifiable data on health and environmental impacts. Therefore, a large effort is required for confirming the data used to reconstruct doses to the public (14). When doses from nuclear facilities were reported they usually were prepared by the operators of the facilities and often were tainted by assumptions and language that conveyed a bias toward minimizing health risks. Moreover, these dose reports were usually based on simplistic models with few estimates of uncertainty. They also lacked peer review of procedures and results and had no public review.

In some cases—particularly for weapons facilities in the United States, Great Britain, France, and the former Soviet Republics—important data documenting health risks to the public and to workers were classified for purported security reasons, and official reports to the public from agencies and scientists with knowledge of the classified material misrepresented health risks to be far lower than they actually were (16–18).

In order for the results of a dose reconstruction to be accepted by the public and the general scientific community, each step should be planned, carried out, and reviewed by members of the public and by independent scientists. Many of the dose reconstructions now underway for Department of Energy (DOE) facilities in the United States are performed with ongoing independent scientific review by panels composed of scientists and representatives of the public. Although some of these dose reconstructions do not involve the public to the greatest extent possible, they are overseen by the public and independent scientists to a degree never before seen. Since dose reconstructions for nuclear weapons facilities may require some data that are still classified, extensive efforts should be made to identify classified material that may be relevant to dose estimates and to declassify these data whenever possible.

In most cases scientists and members of the public involved in recent dose reconstructions for the DOE facilities demand that all data be accessible for review by the public and independent researchers, often requiring extensive declassification efforts.

Risk estimates can be made from reconstructed doses for specific areas around a nuclear facility and for particular periods in time. It is therefore possible to identify groups in the exposed population with the highest risk for disease. Such data are also important for determining the feasibility of epidemiologic studies. An example of this approach is the feasibility analysis performed for the study of thyroid neoplasia in persons exposed to ^{131}I releases from the Hanford nuclear facility (12). In this analysis only preliminary data were available and worst-case conditions were used to estimate statistical power. Because worst-case doses were used, feasibility was evaluated for a range of risks below the maximum, thereby adjusting for overestimates of dose that may erroneously raise statistical power.

The data from this analysis suggested that an epidemiologic study was feasible, and these data were also used in designing an epidemiologic study, which is now underway. The initial worst-case dose estimates also helped to justify a thorough analysis of doses from early Hanford operations—the Hanford Environmental Dose Reconstruction (HEDR) project (*19*). Preliminary dose estimates from the HEDR project helped epidemiologists identify those who were at highest risk for thyroid neoplasia—children who consumed milk from cows that grazed on contaminated pastures.

The question of whether to conduct a dose reconstruction or an epidemiologic study may be one of timing rather than the superiority of one procedure over another. In the United States many federal and state agencies have decided to conduct dose reconstruction before epidemiologic analysis. This decision was made after determining that, in the many instances where epidemiologic studies preceded dose analysis, both additional epidemiologic analysis and dose reconstruction were ultimately required. The relative merits of these two approaches are summarized in Table I.

Dose Reconstruction Projects in the United States

The best examples of state-of-the-art dose reconstructions are those underway at selected DOE facilities (Table II). These studies were initiated in response to public concern over health risks arising from the entire operational histories of the facilities. To assess health risks from these sites, dose reconstructions were chosen instead of epidemiologic studies for two reasons: (1) usually, epidemiologic studies designed and conducted without dose and statistical power estimates were not interpretable; and (2) dose reconstructions provide estimates of health risk that can stand alone or be used to design future epidemiologic studies.

Nevada Test Site. Many efforts were made to estimate radiation doses from the atmospheric weapons tests conducted at the Nevada Test Site (NTS). Most recently the DOE sponsored the Off-site Radiation Exposure Review Project (ORERP), which estimated internal and external exposures for persons who lived near the NTS. The ORERP developed a number of databases with information on fallout deposition. Although the project was monitored by a scientific advi-

Table I. Dose Reconstruction and Epidemiology: A Comparison

Comparison	Dose Reconstruction	Epidemiologic Study
Time		
Identify periods of high and low exposure	Y	N
Estimate exposures for past, current, and future operations	Y	N
Space		
Identify areas impacted by offsite releases	Y	M
Identify areas exposed to different agents	Y	M
Identify pathways for exposure	Y	N
Identify areas in need of cleanup	Y	N
Disease risk		
Rank risks from multiple sources	Y	M
Estimate cumulative risk from exposures	Y	Y
Estimate risks for specific populations	Y	M
Estimate disease rates in study population	N	Y

NOTE: Y is yes; N is no; M is maybe.

sory committee, which included independent scientists, most research was conducted by DOE scientists and contractors and had no public oversight.

The data from the ORERP were used in more comprehensive dose reconstructions conducted in conjunction with epidemiologic studies of childhood leukemia and thyroid disease, sponsored by the National Cancer Institute (20, 21). These studies employed extensive checks on DOE data and rigorous evaluations of models. Estimates of uncertainty were also reported with the doses.

Three Mile Island. Following the accident at the TMI nuclear power facility, many different groups initiated projects to estimate doses received by the public (22). The majority of these studies were conducted in the first year or two after the accident, and four comprehensive reports were prepared by governmental agencies (23–26).

In addition, a thorough and independent dose reconstruction was commissioned by the TMI Fund (3). The atmospheric dispersion data from this reconstruction were used in an epidemiologic study of cancer incidence (4), but actual radiation doses were not employed in this analysis. Because the doses predicted by the model were relatively small, relative atmospheric concentrations predicted by the model were

Table II. Dose Reconstruction Projects in the United States

Facility	Major Exposures	Pathway	Year Started[a] (Sponsor)
Nevada test site	Fission products	Atmosphere, food chain	1979 (DOE)
Three Mile Island	Radioactive noble gases and iodine	Atmosphere	1979 (multiple)
Fernald Feed Materials Production Center	Uranium oxide particles, radon, other radionuclides	Atmosphere, food chain, aquifer, surface water	1985 (DOE) 1990 (CDC)
Hanford	^{131}I, fission products	Atmosphere, food chain, surface water	1988 (DOE) 1991 (CDC)
Rocky Flats	Plutonium, chemicals	Atmosphere, food chain, surface water	1989 (DOE)
Idaho National Engineering Laboratory	Fission products, ^{131}I, plutonium	Atmosphere, food chain, aquifer	1988 (DOE) 1992 (CDC)
Oak Ridge Reservation	Multiple radionuclides and chemicals	Atmosphere, food chain, surface water	1992 [RR, FS] (DOE)
Savannah River site	Multiple radionuclides and chemicals	Atmosphere, food chain, aquifer, surface water	1992 [RR, FS] (CDC)

[a]Projects other than complete dose reconstructions are identified with []: RR = records review; and FS = feasibility study.

used instead of doses. This approach was designed to look for health effects that might be associated with higher releases that were not correctly or honestly reported by official sources and for effects that might be due to synergism between radiation and chemicals released during the accident. This technique is unique and provides a way to conduct exposure–response analysis without detailed knowledge of the quantity of radionuclides released to the environment.

Fernald Feed Materials Production Center. Efforts to estimate off-site radiation doses from the Fernald Feed Materials Production Center (FMPC) began in 1986 in response to a request to the Centers for Disease Control (CDC) for an epidemiologic study of residents around the FMPC. The request was based on evidence of large off-site releases of uranium over the operational history of the plant, which produced uranium oxides and metal ingots from uranium ore concentrates or other uranium-containing materials. The FMPC also processed thorium and stored uranium ore that released radon to the atmosphere around the site.

Instead of conducting an epidemiologic study immediately, the CDC first recommended a dose reconstruction in order to determine the region of potential exposure, to estimate possible radiation doses to the public in this area, and to provide data for determining the feasibility of an epidemiologic study. The DOE, along with staff from the FMPC and other contractors, began a dose reconstruction study in 1986. There were many delays in this project, and there were also many inconsistencies in estimates of the quantities of uranium that were released, as reported in draft versions of the research. In 1989, after discovering a substantial underestimate in the source term for uranium, there was clear evidence that so many errors had been made in the DOE estimates that the final product would not be believed by the public.

In 1990 the CDC contracted with the Radiological Assessment Corporation (RAC) to conduct a thorough dose reconstruction. The completion date is planned for 1994. Draft reports for this project have been released (27–29), and the methodology has been reviewed favorably by a committee of the National Academy of Sciences (30). Oversight has been provided through advisory groups of technical experts and by periodic meetings between RAC scientists, the CDC, and the community around the FMPC.

Hanford. In 1986 the states of Washington and Oregon, with the assistance of the CDC and under the sponsorship of the Nez Perce Tribe, the Yakima Indian Nation, the Confederated Tribes of the Umatilla Indian Reservation, and the Indian Health Service, established the Hanford Health Effects Review Panel to evaluate the possible human health effects associated with the past, current, and future operations of Hanford. The panel was also directed to assess the feasibility and utility of conducting epidemiologic studies around Hanford.

Following a freedom of information request for classified documents about the early operations of Hanford filed by local citizens groups and in compliance with a State of Washington request for the DOE to cooperate with the review of Hanford health risks, several hundred previously classified documents were made public in 1987. Scientists from the State of Washington and the CDC reviewed and summarized these data and made preliminary worst-case estimates of the radiation doses that could have resulted from these exposures. After reviewing these data the Hanford Health Effects Review Panel recommended that more detailed dose estimates be developed for persons living near Hanford during the years of highest radiation release.

In 1988 the DOE established a dose reconstruction project for Hanford—the HEDR project. With the help of representatives of area universities, the DOE selected an oversight panel, which then elected to closely supervise the work of DOE contractors rather than merely provide peer review. In 1991 the CDC assumed responsibilities for funding the HEDR project under its new responsibility for managing and conducting energy-related epidemiologic health research, as specified in a memorandum of understanding between the DOE and the Department of Health and Human Services.

The HEDR project relies heavily on modeling the atmospheric dispersion of all radionuclides released to the environment. It also incorporates analyses of dairy and agricultural practices to estimate doses to the thyroid from ^{131}I and will incorporate data on fish consumption and recreational activities to estimate doses received from radionuclides in the Columbia River. This project is also employing state-of-the-art techniques to model the spatial distribution of doses that could have resulted, as well as to estimate the range of doses received by exposed groups, based on the combined uncertainty of all model parameters. The HEDR project will provide dose estimates from all types of radiation exposure for individual members of the public, based on history of residence and life-style. These data can be used to estimate cancer risks for individuals in the region surrounding Hanford and will be employed in an epidemiologic study of thyroid neoplasia that is being conducted by the CDC and the Fred Hutchinson Cancer Research Center in Seattle, Washington (19).

Rocky Flats. In 1989 the DOE funded the Colorado Department of Health (CDH) to assess health risks around the Rocky Flats facility, which produced plutonium components and other products for nuclear weapons. The CDH relied heavily on the lessons learned from the dose reconstruction at Hanford to establish a multiphased dose reconstruction process with an independent oversight panel. Phase I of the Rocky Flats Toxicologic Review and Dose Reconstruction Project, which was completed in the spring of 1993, was a review of chemicals and radionuclides that possibly were released off-site and a ranking of agents based on risks for cancer in the exposed public (14). Data on source terms and exposed populations were then developed for a detailed dose reconstruction. Phase II of this project, begun in 1993, will involve extensive modeling of releases to the environment, validation of these models with historical and current environmental measurements, and analysis of the uncertainty in dose and risk estimates. These data will also be used to assess the feasibility of epidemiologic studies.

Idaho National Engineering Laboratory. In 1990 the DOE completed a historical dose evaluation of the Idaho National Engineering Laboratory (INEL)—a facility that tested nuclear reactors and other devices and processed spent fuel from the U.S. Navy's nuclear submarines. This analysis was less comprehensive than the dose reconstructions underway at Hanford, Fernald, and Rocky Flats for the following reasons: (1) it did not include chemicals; (2) it did not evaluate groundwater contamination, which is a possible route of exposure; (3) it estimated doses for realistic worst-case conditions but did not provide estimates of uncertainty for these doses; and (4) it was performed by the DOE and its contractors and was not reviewed by the public or independent scientists while it was being conducted.

Because of these deficiencies, which were highlighted by an independent peer review panel convened by the DOE and another independent review panel formed by the state of Idaho, the state requested a more complete dose reconstruction to be performed by the CDC. The CDC began the data collection and evaluation phase of this project in the fall of 1992. At the time of this writing, neither the state of Idaho nor the CDC had provided for adequate public oversight for this project.

Other DOE Facilities. The DOE funded the Tennessee Department of Health to conduct a feasibility study for health-related research at its facilities in Oak Ridge, Tennessee. A review of records that could be used in a dose reconstruction project was started in 1992, overseen by a panel of citizens, scientists, and governmental representatives. Data from this project indicate that a detailed dose reconstruction is technically feasible, and work on this project will begin in 1994.

In 1991 the CDC was given the responsibility for conducting analyses of health effects in populations around all DOE facilities through a memorandum of understanding with the DOE. Under this arrangement the CDC began a similar initial evaluation for the Savannah River site in South Carolina in the fall of 1992. To date neither the state of South Carolina nor the CDC has established ongoing, independent oversight for this project. The CDC is also considering the need for reconstruction efforts around other DOE facilities.

Conclusion

The utility of dose reconstruction for quantifying and explaining health risks will be decided by the success or failure of the studies now underway at Hanford, Fernald, and Rocky Flats. Already, it is clear that dose reconstructions are expensive and time-consuming, ranging in cost between 3 and 10 million dollars or more per site and requiring from 5 to 10 years to complete.

At the conclusion of the aforementioned studies, results will be interpreted by public health experts and the feasibility of epidemiologic studies will be evaluated. Based on the reasons discussed earlier, it is likely that epidemiologic studies will be deemed infeasible at one or more of these facilities.

Although past experience has not shown epidemiologic studies to produce convincing conclusions, it is not yet clear whether the public will be satisfied with risk estimates based solely on data from dose reconstructions. The public may demand epidemiologic studies regardless of their feasibility. Also unclear is whether federal public health agencies will continue to support epidemiologic studies under such conditions, as funds are not available to study every possible disease around every nuclear facility—particularly with well-designed studies that may cost a few million dollars each and take from 5 to 10 years to complete.

What *is* clear is that the dose reconstructions currently underway in the United States are setting new standards for conducting environmental health research, with regard both to improving analytical techniques and to involving the public and independent scientists in the research process. Hopefully, the advances being made in these dose reconstructions will be applied to other areas of environmental health, such as risk assessment for Superfund sites.

It is encouraging that epidemiologists and public health policy makers recognize the usefulness of dose reconstruction and that they seem to agree that these studies are necessary regardless of whether epidemiologic studies are initiated. The value of dose reconstruction will also be recognized at the conclusion of any epidemiologic study that is performed, as that study's results will have to be interpreted for a wider population than those selected for study, and only by using data from a dose reconstruction will such interpretation be possible.

References

1. Shleien, B.; Ruttenber, A. J.; Sage, M. *Health Phys.* **1991**, *61*, 699–713.
2. Hill, A. B. *Proc. R. Soc. Med.* **1965**, *58*, 295.
3. Beyea, J.; DeCicco, J. *Re-estimating the Noble Gas Releases from the Three Mile Island Accident*; Three Mile Island Public Health Fund: Philadelphia, PA, 1992.
4. Hatch, M. C.; Beyea, J.; Nieves, J. W.; Susser, M. *Am. J. Epidemiol.* **1990**, *132*, 397–417.
5. Pacific Northwest Laboratory; *Draft Summary Report, Phase I of the Hanford Environmental Dose Reconstruction Project* (PNL-7410 HEDR, UC-707); Richland, WA, 1990.
6. Richmond, M. C.; Walters, W. H. *Estimates of Columbia River Radionuclide Concentrations: Data for Phase I Dose Calculations* (PNL-7248 HEDR, UC-707); Richland, WA, 1991.
7. Hanf, R. W.; Dirkes, R. L.; Duncan, J. P. *Radioactive Contamination of Fish, Shellfish, and Waterfowl Exposed to Hanford Effluents: Annual Summaries, 1945–1972* (PNWD-1986 HEDR, UC-707); Battelle Pacific Northwest Laboratories: Richland, WA, 1991.
8. Committee on the Assessment of Health Consequences in Exposed Populations, U.S. Department of Energy; *Health and Environmental Consequences of the Chernobyl Nuclear Power Plant Accident*; U.S. Department of Energy: Washington, DC, 1987.
9. Hatch, M. C.; Wallenstern, S.; Beyea, J.; Nieves, J. W.; Susser, M. *Am. J. Publ. Health.* **1991**, *81*, 719–724.
10. Jablon, S.; Hrubec, Z.; Boice, J. D. *JAMA, J. Am. Med. Assoc.* **1991**, *265*, 1403–1408.
11. Rothman, K. J. In *Cancer Epidemiology and Prevention*; Schottenfeld, D.; Fraumeni, J. F., Eds.; W. B. Saunders: Philadelphia, PA, 1982; pp 15–22.
12. Cate, S.; Ruttenber, A. J.; Conklin, A. W. *Health Phys.* **1990**, *59*, 1–10.
13. Beral, V. *Am. J. Epidemiol.* **1990**, *132* (Suppl. 1), S63–S69.
14. Ripple, S. R. *Environ. Sci. Technol.* **1992**, *26*, 1270–1277.
15. Hoffman, F. O.; Gardner, R. H. In *Radiological Assessment: A Textbook on Environmental Dose Analysis*; Till, J. E.; Meyer, H. R., Eds.; Nuclear Regulatory Commission Office of Nuclear Reactor Regulation: Washington, DC, 1983; pp 11-1–11-51.
16. Steele, K. D. *Bull. Atom. Sci.* **1988**, *44*, 17–23.
17. Patterson, W. C. *Bull. Atom. Sci.* **1986**, *42*, 43–45.
18. Shlyakhter, A.; Wilson, R. *Nature (London)* **1991**, *350*, 25.
19. Gilbert, R. O.; Simpson, J. C.; Napier, B. A.; Haerer, H. A.; Liebetrau, A. M.; Ruttenber, A. J.; Davis, S. *Radiat. Res.* **1990**, *124*, 354–355.
20. Stevens, W.; Thomas, D. C.; Lyon, J. L.; Till, J. E.; Kerber, R. A.; Simon, S. L.; Lloyd, R. D.; Abd Elghany, N.; Preston-Martin, S. *JAMA, J. Am. Med. Assoc.* **1991**, *264*, 585–591.

21. Kerber, R. A.; Till, J. E.; Simon, S. L.; Lyon, J. L.; Thomas, D. C.; Preston-Martin, S.; Rallison, M. L.; Lloyd, R. D.; Stevens, W. *JAMA, J. Am. Med. Assoc.* **1993**, *270*, 2076–2082.

22. Beyea, J. *A Review of Dose Assessments of Three Mile Island and Recommendations for Future Research*; Three Mile Island Public Health Fund: Philadelphia, PA, 1984.

23. Auxier, J. A. et al. *Report of the Task Group on Health Physics and Dosimetry to the President's Commission on the Accident at Three Mile Island*; The Kemeny Commission: Washington, DC, 1979.

24. Rogovin, M.; Frampton, G. *Three Mile Island: A Report to the Commissioners and to the Public*; U.S. Nuclear Regulatory Commission Special Inquiry Group: Washington, DC, undated.

25. U.S. Nuclear Regulatory Commission; *Investigation into the March 28, 1979 Three Mile Island Accident* (NUREG-0600); Washington, DC, 1979.

26. Ad Hoc Dose Assessment Group; *Population Dose and Health Impact of the Accident at the Three Mile Island Nuclear Station* (NUREG-0588); U.S. Nuclear Regulatory Commission: Washington, DC, 1979.

27. RAC (Radiological Assessments Corporation); *Task 1: Identification of Release Points at the Feed Materials Production Center*; Report prepared for Fernald Dosimetry Reconstruction Project; RAC: Neeses, SC, 1991.

28. RAC (Radiological Assessments Corporation); *Fernald Dosimetry Reconstruction Project Tasks 2 and 3: Radionuclide Source Terms and Uncertainties*; RAC Report CDC-5; RAC: Neeses, SC, 1993.

29. RAC (Radiological Assessments Corporation); *Fernald Dosimetry Reconstruction Project Task 4: Environmental Pathways—Models and Validation*; RAC Report CDC-3; RAC: Neeses, SC, 1993.

30. Committee on an Assessment of CDC Radiation Studies; *Report of the Committee on an Assessment of CDC Radiation Studies: Dose Reconstruction for the Fernald Nuclear Facility*; National Academy Press: Washington, DC, 1992.

RECEIVED for review October 2, 1992. ACCEPTED revised manuscript April 12, 1993.

Health Effects on Populations Exposed to Low-Level Radiation in China

Lüxin Wei

Laboratory of Industrial Hygiene, Ministry of Public Health, 2 Xinkang Street, P.O. Box 8018, Beijing 100088, China

Several large-scale Chinese investigations on radiation epidemiology are overviewed, and a long-term cancer mortality study in high-background radiation areas (HBRA) is described. Some site-specific cancer incidences were higher among X-ray workers, and a higher mortality rate caused by lung cancer was found among uranium and tin miners who started their work before 1970. An increase in cancer mortality rate was not found among workers in nuclear facilities and inhabitants who lived near the nuclear test site. A cancer mortality study was begun in 1972 in the HBRA whose radiation levels are about three times that of the nearby control areas (CA). About 1 million person-years in HBRA and as many in CA were observed; no increase of site-specific cancer mortalities was found except the cancer of cervix uteri. Data for environmental carcinogens other than natural radiation and data on leukemia and thyroid nodularity were analyzed.

Overview of Several Large-Scale Epidemiological Investigations

Recently, Wang et al. (*1*) reported their long-term investigation on the incidence of malignant tumors among Chinese medical diagnostic X-ray workers in 1950–1985. In a retrospective study, 27,011 X-ray workers from 24 provinces or large cities in China were observed with a control group of 25,782 medical workers other than radiologists. The authors found that the incidences of leukemia, skin cancer, cancer of the esophagus, and cancer of the liver were statistically greater in the study group than in the control group. A higher incidence of leukemia was seen in X-ray workers who started their work before 1970, when radiological protection practices were less stringent. The incidence decreased afterward with the improvement in radiological protection. Skin cancers were few; however, in most cases (six of nine), they occurred on the hands and in eight of nine cases were in those who started their radiological work before 1965. Concerning cancer of the esophagus and the liver, the authors noted that, because the cases were few and other factors that might influence the incidence of these cancers were not investigated, it was difficult to relate these higher incidences to ionizing radiation exposure (Table I).

Finally, the authors emphasized the problem they encountered in their investigation: it was difficult to get accurate dose estimations for these X-ray workers who started their work more than 20 years ago. Further work on dose assessments is needed in the ongoing investigation.

Another type of long-term epidemiological investigation on occupational exposure to ionizing radiation was that related to radon daughter exposure and lung cancer in uranium mines. Yuan et al. (2) investigated a uranium mine in the Hunan Province and reported the results.

A population of 2149 underground uranium miners who tunneled and excavated and started work before 1971 (and who worked for at least 1 year) was observed. Up to 1985 11 of these miners died of lung cancer [person-years (pyr) observed: 31,305] with a crude mor-

Table I. Relative Risk (RR) Estimate of Malignant Neoplasms for X-ray Workers Who Started Work in Different Years

Site or Disease	Before 1959 No. of Cases	RR	1960–1969 No. of Cases	RR	1970–1985 No. of Cases	RR
All	193	1.3	90	1.3	49	0.8
Esophagus	13	5.5	4	5.3	2	3.8
Stomach	26	1.1	7	0.7	3	0.2
Liver	37	1.8	20	2.2	8	1.2
Lung	23	0.7	16	1.6	6	0.7
Skin	7	2.9	1	1.6	1	4.8
Breast	10	1.7	6	1.3	4	1.3
Thyroid	5	2.4	2	1.2	1	1.0
Leukemia	17	2.6	13	3.0	4	1.3

tality rate of 35.14 per 100,000 pyr. Compared with the average rate of middle-size cities of China in 1973–1975 (3), a standard mortality ratio (SMR) of 3.17 was estimated (age adjusted). All the miners with lung cancer died before 1963. The average interval from the beginning of underground work to the occurrence of lung cancer clinically was 18.5 years. The distribution of accumulated exposure (WLM) of the miners and the SMR are shown in Table II.

The authors emphasized that, since the person-years observed were not numerous and the lung cancer cases were few, the confidence intervals of SMR were very wide, especially those in the categories with WLM below 70.

A greater incidence of lung cancer was also found in nonuranium miners. However, the etiology is complicated because of the existence of both radon daughters and other carcinogens in nonuranium underground mines. Since the early 1970s several researchers or research groups investigated the cumulative exposure of tin miners to the radon daughters and the contamination of arsenic-containing ore dust in the underground mines. In the meantime they studied the mortality rates of the miners with lung cancer. Sun et al. (3) reported the accumulated exposure of tin miners to radon daughters and estimated the SMRs of lung cancer between the observed and expected rates (Table III).

The relative risk estimated by Sun was similar to that estimated by other authors for the same tin mines (3–9) but slightly higher than those estimated by the National Institute of Health (NIH, 1985) and by the Committee on Biological Effects of Ionizing Radiation (BEIR-IV, 1988), who analyzed the data from investigations in uranium mines of the United States, Canada, and Czechoslovakia and an iron mine in Sweden (Table IV).

Table II. Distribution of Accumulated Exposure (in WLM) of the Miners and the SMR

Category of WLM[a]	Person-Years Observed	No. of Lung Cancer Deaths Observed	No. of Lung Cancer Deaths Expected	SMR (90% CI)[b]
<15 (8.77)	9958	2	0.78496	2.55 (9.19–0.31)
15–70 (27.06)	16,431	4	1.68953	2.37 (6.06–0.65)
71–200 (125.90)	3800	3	0.76522	3.92 (11.46–0.81)
201–500 (275.04)	1116	2	0.23026	8.69 (31.36–1.05)
Total (43.14)	31,305	11	3.46538	3.17 (5.68–1.58)

[a]Figures in parentheses are the averages; to convert these figures to SI units, multiply WLM by 3.5×10^{-3} J h m^{-3}.
[b]CI = confidence interval.
SOURCE: Data are from reference 2.

Table III. Accumulated Exposure (WLM) and SMRs for Lung Cancer in Tin Miners

Group	Person-Years Observed	WLM	No. of Lung Cancer Deaths		SMR	Relative Risk (% WLM)
			Observed	Expected[a]		
Mines L and M 1974–1978 data[b]	23,995	500	81	8.45	9.59	1.72
Mine L 1960–1984 data[c]	25,506	1000	223	17.80	12.53	1.15

[a]From the lung cancer mortality data of workers and inhabitants above the mines.
[b]Miners who started underground work before 1970.
[c]Miners who started underground work before 1949.
SOURCE: Data are from reference 3.

Table IV. Relative Risk (Percent per WLM) Estimated by Various Authors

Author	Source	Relative Risk	Reference
U.S., NIH (1985)	Uranium miners (U.S., Czechoslovakia)	1.2–1.5	4
U.S., BEIR-IV (1988)	Uranium mines in U.S. and Canada and iron mine in Sweden	1.3–1.5	5
Li et al. (1989)	Geological prospecting teams (for uranium mine)	1.0	6
Mao (1982)	Yunnan tin miners	1.7	7
Wang et al. (1984)	Yunnan tin miners	1.5	8
Sun (1987)	Yunnan tin miners	1.2–1.7	3
Lubin et al. (1990)	Yunnan tin miners	1.7	9

Based on the preceding observations and comparisons, combined with the results from laboratory experiments, Sun suggested that, in the case of the Yunnan tin mines, the high concentration of radon daughters existing before 1970 was the main contributor to the lung cancer of these underground miners; arsenic dust was an additional factor of carcinogenesis (3).

An epidemiological investigation on the detrimental effects of workers in nuclear installations (those who work with fuel element production and fuel reprocessing) was also conducted in China; no increase of cancer mortality was found, as Sun mentioned in his article (10). However, a detailed report concerning this investigation remains unpublished.

Zhou et al. reported the results of their investigation on dose assessments and health surveys in the residential areas around the Xinjiang nuclear test site (11). These areas include the villages and towns near the test site (Lop Lake); the nearest inhabitant point is a small town (M-Town) 120 km away from the test site. About 300,000 inhabitants who lived in these study areas were observed. The controls were selected from the windward areas, partially isolated by the mountains from the test site; much less local fallout deposited on the control areas during and after the nuclear test. About 200,000 inhabitants lived there. The geological structure, industrial and agricultural production, population structure, living conditions, and medical sanitary facilities were comparable to the study areas. The accumulated deposition of ^{90}Sr, ^{137}Cs, and total plutonium on the surface soil (30-cm deep) were measured in uncultivated land. The results are summarized in Table V. The effective dose commitment from the local fallout was estimated to be 58.8 µSv.

Table V. Accumulated Deposition of ^{90}Sr, ^{137}Cs, and Plutonium
on the Surface Soil (to 30-cm Deep) of Uncultivated Land

| Area | Accumulated Activity | | |
	^{90}Sr $(\times 10^3\ Bq/m^{-2})$	^{137}Cs $(\times 10^3\ Bq/m^{-2})$	Total Plutonium (Bq/m^{-2})
Study	1.5 ± 0.3	4.9 ± 4.0	123.6 ± 67.7
Control	1.3 ± 0.2	3.9 ± 0.5	93.4 ± 15.0
M-Town[a]	1.3 ± 0.3	16.7 ± 9.0[b]	321.7 ± 86.0

[a]M-Town—120 km (south west) from test site.
[b]Different significantly.

A retrospective study on cancer mortality (1974–1984 data) in the study areas (542,782 person-years) and control areas (756,314 person-years) was conducted and showed that the overall cancer mortality rate (adjusted with the combined population of study and control areas) was lower in the study areas (55.51 per 100,000 pyr) than it was in the control areas (59.09 per 100,000 pyr), but the difference was not statistically significant. Also, no statistical difference for leukemia mortality between these two areas was found. Nineteen types of hereditary diseases and congenital deformities were examined clinically in children born from 1974 to 1984. The prevalence of these diseases in the study areas (67 of 9408) was similar to that in the control areas (81 of 11,124). Similarly, the incidences of Down's syndrome were 0.6 per thousand in the study areas and 0.54 per thousand in the control areas.

Cancer Mortality Study in High-Background-Radiation Areas of Yangjiang

The High-Background-Radiation Research Group (HBRRG) began its health survey in the high-background-radiation areas (HBRA) of Yangjiang in 1972. The purpose of this investigation is to provide information for evaluating whether any detrimental effects exist in a large population whose families were continuously exposed to higher natural radiation for many generations.

The investigated HBRA are two separated areas that cover a total area of about 500 square kilometers (Figure 1). The sources of higher background radiation are nearby mountains, whose surface rocks are granites from which fine particles of monazite are washed down continuously year after year by rain and are deposited in the surrounding basin regions. The geological structure in these two areas also plays an important role. The control areas (CA) with normal radiation background were selected not far from the HBRA [the closest points were about 10 km away (Figure 1)].

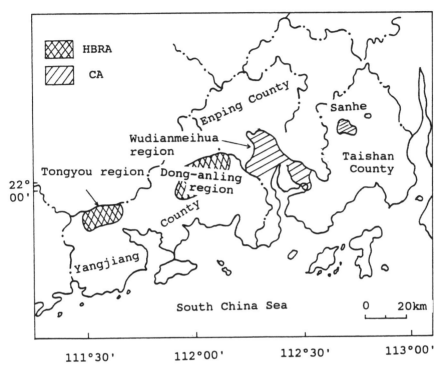

Figure 1.   Locations of the investigated areas.

The results of analyses of soil samples (50-cm deep) by gamma spectrometry and radiochemistry, as well as the measurements of field gamma spectrometry, showed that the concentrations of natural radionuclides in soils of HBRA were significantly higher than those in CA, especially those of thorium, which were about six times higher than those in CA (*12*) (Table VI).

Construction materials of houses in HBRA were also contributors to the higher background of gamma radiation, because they came from the local earth. Table VII shows the differences in exposure rates measured at various locations. Indoor exposure rates were higher than those outside.

For purposes of estimating the doses from external gamma radiation and their distribution among the investigated inhabitants, environmental gamma exposure rates and individual cumulative gamma exposure were measured with various devices and dosimeters. Detailed information concerning the methodology and procedures was published elsewhere (*13, 14*). The annual measured individual doses from gamma exposures in HBRA and CA are shown in Table VIII. A preliminary classification of dose groups (gamma radiation only) and

Table VI. Concentrations of Natural Radionuclides in Soils of Investigated Areas (Bq/kg)

Method of Analysis (Year of Measurement)		^{238}U	^{232}Th	^{226}Ra
HBRA				
Radiochemistry	(1975)	93.5 ± 17.2	177.1 ± 82.2	70.3 ± 27.1
	(1979)	93.5 ± 20.9	248.2 ± 115.1	144.3 ± 55.5
Gamma spectrometry				
of samples	(1979)	118.1 ± 34.4	227.3 ± 85.1	136.9 ± 44.4
	(1982)	109.5 ± 40.6	236.3 ± 130.2	122.1 ± 40.1
Field gamma				
spectrometry	(1982)	87.3 ± 28.3	149.2 ± 52.2	—
Average		100.4 ± 28.3	207.6 ± 87.5	118.4 ± 41.9
CA				
Radiochemistry	(1975)	20.5 ± 9.8	34.9 ± 13.6	22.2 ± 7.4
	(1979)	20.9 ± 8.6	32.5 ± 12.7	29.6 ± 11.1
Gamma spectrometry				
of samples	(1979)	29.5 ± 11.1	36.6 ± 16.8	29.6 ± 11.1
	(1982)	27.1 ± 8.6	36.2 ± 13.6	29.6 ± 25.7
Field gamma				
spectrometry	(1982)	42.6 ± 7.38	27.1 ± 10.7	—
Average		26.3 ± 9.1	33.5 ± 13.5	27.8 ± 13.8
Ratio of HBRA to CA		3.8	6.2	4.3

Table VII. Gamma Radiation Exposure Rates Measured at Various Locations

	Exposure Rate (Average ± SD) (μR/h)			
Area	Surface of Bed	Indoor	Outdoor	Farm Field
HBRA	80.06 ± 13.04	78.14 ± 12.39	49.94 ± 8.87	37.15 ± 8.48
CA	22.26 ± 4.17	21.74 ± 3.77	13.14 ± 7.55	10.36 ± 2.34

NOTE: 2512 and 2245 households were measured in HBRA and CA, respectively. "Indoor" includes sitting rooms and kitchens; "outdoor" includes yards, lanes, and roads.

distribution of person-years observed in these groups in 1979–1986 are shown in Table IX.

For purposes of estimating the internal doses from natural radionuclides, concentrations of ^{222}Rn, ^{220}Rn, and the potential alpha energy of decay products, concentrations of thorium isotopes in human lung tissue, and concentrations of ^{226}Ra and ^{228}Ra in human teeth and bone, activity concentrations of ^{210}Po and ^{210}Pb in human tissues were measured (15–19). The results are shown in Tables X through XIV.

Based on the preceding data and those from the investigation on intakes of natural radionuclides from food and drinking water (19) and using the calculation models of the United Nations Scientific Committee on the Effects of Atomic Radiation [UNSCEAR (20)], the HBRRG estimated the annual effective doses from natural radiation sources in HBRA and CA (Table XV).

Table VIII. Estimate of Annual Individual Dose from External Gamma Exposure (10^{-5} Gy/a) Based on Results of Various Measurement Techniques

Device or Dosimeters	Fixed Point Measurements (Average)		Dosimeters Worn by People (Average)	
	HBRA	CA	HBRA	CA
RSS-111	211.6	77.6		
FD-71	222.1	79.0		
RPL	203.8	79.8	208.0	79.7
TLD				
$CaSO_4$:Dy	196.1	70.8	208.0	79.7
$CaSO_4$:Tm			220.7	80.5
LiF:Mg,Ti			208.0	70.0

NOTE: Average of total data: 2.10 mGy/a for HBRA; 0.77 mGy/a for CA.

Table IX. Preliminary Classification of Dose Groups[a] and Distribution of Person-Years Observed in 1979–1986

Group	Area	Dose Range (mGy)	Dose Weighted Average (mGy)	Sex	No. of Person-Years	Percentage[b]
1	CA	0.58–1.03	0.75	M	298,526	100
				F	289,299	100
2	HBRA	1.55–2.00	1.85	M	108,087	32.8
				F	97,432	33.1
3	HBRA	2.01–2.40	2.15	M	153,094	46.5
				F	138,529	47.1
4	HBRA	2.41–3.10	2.55	M	68,353	20.7
				F	58,076	19.8

[a]Data analyzed and calculated by Pan (1990).
[b]Percentage either for male or female in HBRA or CA.

The cumulative doses received by individuals will increase with the increase in their ages. Fifty-year-old people would receive a cumulative dose from natural gamma radiation in HBRA of about 105 mGy (87–129 mGy) and an effective dose of 274 mSv (248–369 mSv).

The survey of demographic data has been followed up since 1975 with the help of local governments. The investigated areas were rural; about 94 and 93% of the total inhabitants were farmers (peasants) in the HBRA and CA, respectively (22). The population structure and male/female ratios of HBRA and CA are shown in Table XVI. As evidenced by the data the investigated populations were younger than the standard population of the world and those in European countries (Table XVII reference 23). The difference of age structure may influence the mortality rates of cancer in different population cohorts.

Table X. Concentrations of Radon and Decay Products in Air of Investigated Areas

Area	No. of People in Places Measurements Taken	Places Measurements Taken	Concentration of ^{222}Rn (Bq m^{-3})	Concentration of ^{220}Rn (Bq m^{-3})	Alpha Potential Energy of ^{222}Rn Decay Products		Alpha Potential Energy of ^{212}Pb + ^{212}Bi	
					WL	Jm^{-3}	WL	Jm^{-3}
HBRA	35,021	Indoors	31.8	167.5	0.00482	0.100×10^{-6}	0.01227	0.255×10^{-6}
CA	39,301	Outdoors	16.4	18.4	0.00464	0.097×10^{-6}	0.00255	0.053×10^{-6}
		Indoors	11.1	17.5	0.00181	0.038×10^{-6}	0.00334	0.069×10^{-6}
		Outdoors	11.2	3.8	0.00212	0.044×10^{-6}	0.00097	0.020×10^{-6}

Table XI. Concentrations of Thorium Isotopes in Human Lung Tissue
in HBRA and CA (mBq/kg Fresh Tissue)

Area	No. of Subjects (Autopsy)	Concentration (Mean ± SD)		
		^{232}Th	^{230}Th	^{228}Th
HBRA				
Age < 60 yr	0			
Age > 60 yr	4	222.37 ± 218.67	227.55 ± 204.98	330.41 ± 332.26
Total	4	222.37 ± 218.67	227.55 ± 204.98	330.41 ± 332.26
CA				
Age < 60 yr	4	40.70 ± 19.24	30.34 ± 14.06	46.62 ± 34.41
Age > 60 yr	4	35.15 ± 10.36	28.49 ± 2.22	35.15 ± 2.22
Total	8	38.11 ± 14.80	29.23 ± 9.25	40.70 ± 23.31
HBRA/CA		5.84	7.79	8.12

Table XII. Specific Activities of ^{226}Ra and ^{228}Ra in Human Teeth and
Their Calculated Contents in Bones

Area	No. of Teeth Measured	Specific Activity (Bq/kg Tooth Ash)		Deduced Contents (Bq/kg Bone)	
		^{226}Ra	^{228}Ra	^{226}Ra	^{228}Ra
HBRA	1100	2.78 ± 0.19	1.96 ± 0.33	1.52 ± 0.10	1.07 ± 0.18
CA	900	0.81 ± 0.33	0.56 ± 0.11	0.44 ± 0.18	0.31 ± 0.06
HBRA/CA		3.4	3.5	3.4	3.5

Table XIII. Activity Concentrations of ^{226}Ra and ^{228}Ra in Human Bones

No. of Cases (Autopsy)		Specific Activity (Bq/kg Ash)		Content (Bq/kg Bone)	
		^{226}Ra	^{228}Ra	^{226}Ra	^{228}Ra
HBRA	10	2.58 ± 0.51	2.16 ± 0.64	1.44 ± 0.29	1.21 ± 0.36
CA	8	0.75 ± 0.45	0.77 ± 0.44	0.42 ± 0.05	0.43 ± 0.25
HBRA/CA		3.4	2.8	3.4	2.8

Table XIV. Activity Concentrations of ^{210}Po and ^{210}Pb in Human Tissues
(Autopsy)

Area	Tissue	No. of Cases	^{210}Po (Bq/kg)		^{210}Pb (Bq/kg)	
			Range	Average	Range	Average
HBRA	Lung	2	0.71–0.82	0.75	1.07–1.37	1.24
	Liver	2	2.11–2.55	2.35	0.63–1.67	1.15
	Rib	1	7.07	7.07	7.27	7.27
CA	Lung	8	0.07–0.33	0.21	0.09–0.51	0.31
	Liver	4	0.41–0.70	0.60	0.31–0.40	0.37
	Rib	3	2.52–3.37	2.93	2.78–3.55	3.20

Table XV. Annual Effective Doses Resulting from Natural Radiation Sources in HBRA and CA

Sources	Annual Effective Dose (mSv)	
	HBRA	CA[a]
External irradiation		
Terrestrial gamma radiation	1.85	0.52 (0.41)
Cosmic rays		
Ionizing components	0.23	0.23 (0.30)
Neutron components	0.02	0.02 (0.05)
Subtotal	2.10	0.77 (0.80)
Internal irradiation		
^{40}K	0.18	0.18 (0.18)
^{87}Rb	0.006	0.006 (0.006)
^{226}Ra	0.087	0.027 (0.07)
^{222}Rn	0.03	0.009
Rn's decay products	2.320[b]	0.960[b] (1.100)
^{228}Ra	0.195	0.058 (0.013)
^{220}Rn + ^{216}Po	0.095	0.011
^{212}Pb + ^{212}Bi	1.360[b]	0.400[b] (0.16)
Subtotal	4.273	1.651 (1.589)
Total (rounded)	6.4	2.4 (2.4)

[a]Figures in parentheses are the world average values reported by UNSCEAR (20).
[b]Figures re-estimated using the models of the National Council on Radiation Protection and Measurements (21), which are higher than our previous estimates. However, the original data are the same.

However, the population structure in HBRA was similar to that in CA (24).

The HBRRG investigated not only the population structure, which may influence the assessment of the carcinogenic effect of radiation, but also the factors believed to affect the occurrence of mutation-based diseases, either those environmentally caused or those of the host. The results obtained from a case-control study using two-stage sampling are shown in Table XVIII. Generally, these factors in CA seemed comparable to those in HBRA (25, 26).

The studies of cancer mortalities in HBRA and CA were begun in 1972. The early data were obtained by means of retrospective surveys (1972 for the start of pilot study, 1975 for the 1970–1974 data, and 1979 for the 1975–1978 data). A cancer registry system was established in 1979 for the investigated areas; in this system local physicians, with the help of section and county hospitals and many health administrative organizations, report all cancer cases and cancer-related deaths. Diagnoses are checked, sometimes re-examined, and confirmed by an expert group who meet twice a year at the investigated areas to evaluate cases. Meanwhile, deaths from all causes are also registered and analyzed. By the end of 1986, 467 cancer deaths were

Table XVI. Male/Female Ratio[a] of the Investigated Populations in HBRA and CA (1970–1985)

	HBRA			CA		
Age	Male (Pyr)	Female (Pyr)	Ratio (M/F)	Male (Pyr)	Female (Pyr)	Ratio (M/F)
5	124,063	110,312	1.12	96,294	90,794	1.06
15	120,578	106,736	1.13	112,106	102,288	1.10
25	79,020	65,256	1.21	89,837	79,416	1.13
35	52,745	38,115	1.38	54,114	43,458	1.25
45	43,213	37,770	1.14	40,611	38,918	1.04
55	32,119	33,853	0.95	35,519	41,360	0.86
65	20,845	22,678	0.92	21,984	28,766	0.76
75	9009	13,702	0.66	9091	16,926	0.54
85	1567	3624	0.43	1721	4416	0.39
Total	483,159	432,046	1.12	461,274	446,342	1.03

[a]Male/Female ratio: person-years for males/person-years for females.

Table XVII. Age Structure (Extract) of Inhabitants in HBRA[a], CA[a], and Two WHO[b] "Standard Populations"

	Percentage			
Age (in Years)	HBRA	CA	World	European
0–29	66.3	62.9	56.0	43.0
30–69	30.6	33.6	40.0	50.0
70 and above	3.1	3.5	4.0	7.0
Total	100.0	100.0	100.0	100.0

[a]Data averaged from 1970–1986 statistics.
[b]SOURCE: Data are from reference 25.

Table XVIII. Comparison of Factors Known to Affect Diseases Caused by Mutation Processes

	Constituent Ratio (%)			Odds Ratio[a]	
Factors	HBRA	CA	P[b]	Matched Analysis	Unmatched Analysis
Pesticide use	62.3	63.6	NS	1.00	0.95 (0.72–1.24)
Occupations involving the use of poisonous and noxious substances	1.5	2.0	NS	0.60 (0.15–2.47)	0.77 (0.29–2.09)
Smoking	37.9	37.6	NS	1.06 (0.71–1.58)	1.01 (0.88–1.16)
Alcohol consumption	37.2	38.6	NS	0.85 (0.60–1.21)	0.90 (0.64–1.27)
Medical X-ray exposure	20.0	26.4	NS	0.62 (0.45–0.87)	0.70 (0.51–0.95)

[a]Figures in parentheses represent 95% confidence intervals.
[b]NS: no significance.

Table XIX. Crude Cancer Mortality Rates in HBRA and CA (1970–1986, per 100,000 Person-Years)

Sex	Area	Person-Years Observed	Crude Mortality Rate (95% Confidence Interval)	Rate Ratio[a] (95% Confidence Interval)
Male	HBRA	530,952	56.3 (50.1–63.1)	0.93 (0.81–1.06)
	CA	504,458	65.4 (58.6–77.9)	$P = 0.65$
Female	HBRA	477,817	35.2 (30.1–41.0)	0.96 (0.81–1.14)
	CA	490,612	41.4 (35.9–47.5)	$P = 0.35$

[a]Age adjusted by combined population of HBRA and CA.

Table XX. Order of Ten Site-Specific Cancer Mortality Rates in HBRA and CA (Male and Female Data Combined)[a]

	HBRA		CA	
Order	Site	Rate[b] (per 100,000 pyr)	Site	Rate[b] (per 100,000 pyr)
1	Liver	12.50	Liver	13.92
2	Nasopharynx	9.81	Nasopharynx	10.45
3	Stomach	5.60	Stomach	4.45
4	Bone-marrow (Leukemia)	3.02	Bone-marrow (Leukemia)	3.39
5	Lung	2.65	Lung	3.29
6	Intestine	1.70	Intestine	2.28
7	Esophagus	1.40	Esophagus	1.49
8	Cervix, uterus	1.37	Breast	1.05
9	Breast	0.75	Bone sarcoma	0.59
10	Bone sarcoma	0.52	Cervix, uterus	0.45

SOURCE: Adapted with permission from reference 28. Copyright 1988.
[a]In HBRA, 1,008,769 person-years, in CA, 995,070 person-years were observed.
[b]Adjusted with the combined population of HBRA and CA.

Table XXI. Order of Nine Site-Specific Cancer Mortality Rates in HBRA and CA (Males)

	HBRA		CA	
Order	Site or Disease	Rate[a] (per 100,000 pyr)	Site or Disease	Rate[a] (per 100,000 pyr)
1	Liver	16.67	Liver	21.62
2	Nasopharynx	11.58	Nasopharynx	13.79
3	Stomach	7.11	Stomach	5.49
4	Lung	3.36	Lung	3.39
5	Leukemia	3.21	Leukemia	3.70
6	Intestine	1.96	Intestine	2.69
7	Esophagus	1.92	Esophagus	1.32
8	Osteosarcoma	0.78	Osteosarcoma	0.77
9	Breast	0.00	Breast	0.00

SOURCE: Adapted with permission from reference 28. Copyright 1988.
[a]Adjusted for the combined population of HBRA and CA.

Table XXII. Order of Ten Site-Specific Cancer Mortality Rates in HBRA and CA (Females)

	HBRA		CA	
Order	Site or Disease	Rate[a] (per 100,000 pyr)	Site or Disease	Rate[a] (per 100,000 pyr)
1	Nasopharynx	7.82	Nasopharynx	7.08
2	Liver	6.67	Liver	6.14
3	Stomach	3.82	Stomach	3.42
4	Cervix, uterus	2.94	Lung	3.23
5	Leukemia	2.80	Leukemia	3.06
6	Lung	1.82	Breast	2.51
7	Breast	1.60	Intestine	2.09
8	Intestine	1.37	Esophagus	1.68
9	Esophagus	0.68	Cervix, uterus	0.94
10	Osteosarcoma	0.21	Osteosarcoma	0.40

SOURCE: Adapted with permission from reference 28. Copyright 1988.
[a]Adjusted for the combined population of HBRA and CA.

found among 1,008,769 person-years at risk in HBRA, resulting in a crude mortality rate of 46.29/100,000 pyr. The corresponding figures in CA were 533 cancer deaths, 995,070 pyr, and a crude cancer mortality rate of 53.56/100,000 pyr. Mortality rates from all cancers are shown in Table XIX (27). Orders of site-specific cancer mortalities in HBRA and CA for both sexes are shown in Tables XX through XXII.

From the preceding data it can be seen that the mortality rate for all cancers was higher in CA than that in HBRA for males, females, and for the combined sexes. However, the differences were not statistically significant (28). For site-specific cancers only small differences existed between HBRA and CA; however, mortality rates of cervix uteri cancer were found to be significantly different between these two areas. The induction of this type of cancer though is difficult to relate to ionizing radiation exposure.

Further analyses of these data by the HBRRG follows.

1. As the data of age-specific mortalities shows, most cancer cases appeared in the 40 years and older age group; these data are consistent with those appearing spontaneously.

   For these reasons the Research Group analyzed the cancer mortality of all cancers except leukemia of HBRA and CA inhabitants aged 40–70 years (27). The results (Table XXIII) were somewhat different from those of analyses of cancer mortalities for all age groups. The mortality rate in HBRA was significantly lower, although there is a wide confidence interval. Accumulation of more person-years for these older age groups is necessary.

Table XXIII. Mortality Rates of All Cancers Except Leukemia of HBRA and CA Inhabitants Aged 40–70 Years (1970–1986)[a]

Area	Person-Years Observed	Mortality Rate $(10^{-5})^b$	β Value[a] (95% CI)	P Value
HBRA	207,900	143.8 (299)	−14.6%	
			(−24.8, −3.0%)	0.04
CA	224,380	168.0 (377)		

[a]The computer program "AMFIT" (Preston, 1987) was used to fit a Poisson regression model. R_{HB} (S, T) = R_{CA} (S, T)(1 + β). R_{HB} and R_{CA} are the mortality rates in HBRA and CA, respectively. S is sex; T is age; and β is the "excess" rate of HBRA to CA.
[b]Figures in parentheses are numbers of cancer deaths.

Table XXIV. Comparison of Excess Relative Risk of Cancer Mortality in Different Dose Groups of HBRA[a]

Dose Group (mGy/year)	Sex	β Value	Standard Errors	P Value[b]
1.85	M	−0.114	0.135	NS
	F	0.045	0.194	NS
2.15	M	0.004	0.132	NS
	F	0.015	0.163	NS
2.55	M	0.001	0.170	NS
	F	−0.154	0.208	NS

[a]From 1979–1986 data only.
[b]β values for these three groups were not different statistically.

Table XXV. Mortality Rates of Malignant Neoplasms and Leukemia in Hong Kong and Some Asian Countries or Areas (per 100,000 Population)

Country or Area	Malignant Neoplasms (All Ages)		Leukemia (All Ages)	
	Male	Female	Male	Female
Hong Kong (1986)[a]	172.7	116.7	3.4	3.3
Japan (1986)[a]	191.1	126.9	5.2	3.7
South Korea (1985)[a]	95.0	54.7	2.9	2.4
Singapore (1986)[a]	127.5	95.2	3.1	3.2
Sri Lanka (1982)[a]	25.2	24.0	—	—
China (1977)[b]	84.35	63.16	2.8	2.24
	(119.6)	(80.7)		
HBRA[c]	56.31	35.16	3.20	2.93
CA[c]	65.42	41.38	3.57	3.06

[a]Data are from reference 23.
[b]Data are from reference 29. Figures in parentheses are world standardized mortality rates.
[c]Crude mortality rate.

2. To identify whether a dose–effect relationship exists in the HBRA, an internal comparison of excess relative risks of cancer mortality (β-value) was made between the different dose groups in HBRA (Table XXIV). The result showed that there was no statistical difference found. One possibility is that a practical threshold exists in such low-level radiation exposure. Another explanation is that the size of person-years of the observation is not large enough to identify the minor changes.

3. Leukemia was recognized by the scientific community as a malignancy closely related to ionizing radiation exposure, and it has a shorter latency period than do solid tumors. In addition, there are fewer missed cases and misdiagnoses in the cancer registry.

   Data published by the World Health Organization (23) indicated that, although cancer mortality rates fluctuated widely in various countries or areas, the mortality rates of leukemia in Asian countries or areas closely agreed. The mortality rate of leukemia in HBRA was found to be within the range of the spontaneous incidence in CA and near that in China (Table XXV, references 27 and 29).

4. Since the thyroid nodularity may be a kind of predisposition to thyroid cancer, the HBRRG conducted a collaborative study with the National Cancer Institute of the United States to identify if high-background radiation would produce a detectable increase in thyroid nodularity. Because the female's thyroid is more sensitive to radiation and older people receive larger cumulative doses, women aged 50–65 ($N = 1001$ in HBRA and 1005 in CA) who resided in HBRA or CA throughout their entire lives were selected. Personal interviews and physical examinations by experienced physicians of thyroid diseases were conducted on all women; thyroid hormone levels, urinary iodine, and chromosomal aberrations were measured and analyzed for some women, randomly selected. No evidence that nodular thyroid disease was elevated among women in HBRA compared to that in CA was found. The prevalence of nodular thyroid diseases were 9.5% in HBRA and 9.3% in CA (30).

   These data as a whole suggest that continuous exposure to low-level radiation of the HBRA is unlikely to appreciably increase the risk of nodular thyroid disease.

Table XXVI. A Comparison of Excess Relative Risk Estimates (per 10 mSv) With 90% Confidence Limits for All Cancers Except Leukemia

Group	Excess Relative Risk per 10 mGy with 90% Confidence Limits for All Cancer Except Leukemia	
Inhabitants in HBRA, Yangjiang, China	−0.71%	(<0, 0.31%)
Atomic bomb survivors, Hiroshima and Nagasaki[a]	0.41%	(0.32%, 0.51%)

[a]As presented in Shimizu et al. (*31*). The original values were given as risk per Gray.

Preliminary statistical analyses of the risk of excess cancer deaths were conducted from the data of cancer mortalities in HBRA and CA. The excess relative risk estimate (per 10 mSv) with 90% confidence limits for all cancer except leukemia is shown in Table XXVI. A comparison with the Radiation Effects Research Foundation (RERF) estimate is also shown in the table. The exposure patterns these two cohorts were different. Atomic-bomb survivors received acute high dose rate irradiation, while HBRA inhabitants received continuous low dose rate exposure. The original risk values estimated by the RERF were expressed in Gray units (*31*).

Acknowledgments

The author wishes to thank Wang Jianzhi and Pan Yingdong of the Laboratory of Industrial Hygiene, Ministry of Public Health, China, for their statistical analyses. The author is also grateful to Zhang Jihui of the same laboratory for her careful word processing of the original manuscript of this chapter.

References

1. Wang, Jixian; Inskip, P. D.; Boice, J. D., Jr.; Li, Benxiao; Zhang, Jingyuan; Fraumeni, J. F., Jr. (compilers) *Chin. J. Radiol. Med. Prot.* **1991**, *11*(3), 149–154.
2. Yuan, Liyun; Gao, Yan; Gu, Juanjuan; Wang, W.; Jiang, L.; Zhang, D.; Han, B. *Chin. J. Radiol. Med. Prot.* **1990**, *10*(4), 230–234.
3. Sun, Shiquan; You, Zhanyun *Chin. J. Radiol. Med. Prot.* **1987**, *7*(4), 225–229.
4. Report of the National Institute of Health Ad Hoc Working Group to Develop Radioepidemiological Tables; NIH Publication 85-2748; U.S. Government Printing Office: Washington, DC, 1985.
5. National Research Council, Committee on the Biological Effects of Ionizing Radiations; *Health Risks of Radon and Other Internally Deposited Alpha-Emitters (BEIR-IV)*; National Academy Press: Washington, DC, 1988.

6. Li, Suyun; Zhang, Shenghui; Bao, Shoushen; Hou, H.; Meng, X.; Lai, C.; Wang, J. *Fushe Fanghu* **1989**, 9(4), 289–294.
7. Mao, Baolin *Oncology* **1982**, 2, 1.
8. Wang, Xianhua; Huang, Xinghui; Huang, Songlin; Zhang, X.; Zhang, J.; Zhang, F.; Mao, B.; Yang, S.; Pan, Z. *Chin. J. Radiol. Med. Prot.* **1984**, 4(3), 10–14.
9. Lubin, J. H.; Jay, H.; Qian, Y.; Taylor, P. R.; Yao, S.; Schatzkin, A.; Mao, B.; Rao, J.; Xuan, X.; Li, J. *Cancer Res.* **1990**, 50, 174.
10. Sun, Shiquan. *Chin. Radiol. Med. Prot.* **1989**, 9(5), 320–324.
11. Zhou, Wenliang; Wei, Jiguan; Zhang, Jujing; Xu, H.; Yang, Y.; Quo, H. *Chin. Radiol. Med. Prot.* **1989**, 9(2), 87–91.
12. Luo, Daling; Zhang, Chunxiang; Guan, Zujie; Lai, X.; Li, Z.; Weng, S. *Chin. J. Radiol. Med. Prot.* **1985**, 5(2), 90–93.
13. Yuan, Yongling; Shen, Hong; Zhou, Zhixin; Huang, G.; Zeng, X.; Fan, J.; Hu, Z. *Chin. J. Radiol. Med. Prot.* **1982**, 2(2), 16.
14. He, Miaoting; Cui, Guangzhi *Chin. J. Radiol. Med. Prot.* **1985**, 5(2): 87–90.
15. Zhang, Zhonghou *Chin. J. Radiol. Med. Prot.* **1985**, 5(2), 102–104.
16. Wei, Luxin; Zha, Yongru; Tao, Zufan; He, W.; Chen, D.; Yuan, Y. In *Epidemiological Investigations on the Health-Effects of Ionizing Radiation*; Institut für Strahlenschutz, der Berufsgenossenschaft der Fein Mechanik und Electrotechnic und der Berufsgenossenschaft der Chemischen Industrie: Cologne, Germany, 1988; pp 7–25.
17. Zhao, Jianhua; Wang, Jufen; Xu, Ning; Hu, Y.; He, W.; Huang, L.; Chen, L.; Li, R.; Ye, H. *Chin. J. Radiol. Med. Prot.* **1982**, 2(2), 38–40.
18. Zhao, Jinghua; Tang, Yaoyuan *Chin. J. Radiol. Med. Prot.* **1985**, 5(2), 97–99.
19. High Background Radiation Research Group; *Radiol. Med. Prot.* **1978**, 2, 27–51.
20. United Nations Scientific Committee on the Effects of Atomic Radiation; *1988 Report to the General Assembly, Annex A: Exposures from Natural Sources of Radiation*; United Nations: New York, 1988.
21. National Council on Radiation Protection and Measurements; *Report No. 93: Ionizing Radiation Exposure of the Population of the United States*; NCRP: Bethesda, MD, 1987.
22. Tao, Zufan; Wei, Luxin *J. Radiat. Res.* **1986**, 27, 141–150.
23. World Health Organization; *1987 World Health Statistics Annual*; WHO: Geneva, Switzerland, 1987.
24. Wei, Luxin; *Health Phys.* **1988**, 54(2), 222–224.
25. High Background Radiation Research Group; *J. Radiol. Med. Prot.* **1985**, 5, 82–87.
26. Tao, Zufan; Li, Hong; Zha, Yongru; Lin, Z. *Chin. J. Radiol. Med. Prot.* **1985**, 5, 130–135.
27. Wei, Luxin; Zha, Yongru; Tao, Zufan; He, W.; Chen, D.; Yuan, Y. *J. Radiat. Res.* **1990**, 31, 119–136.
28. Wei, Luxin; Zha, Yongru; Tao, Zufan; He, W.; Chen, D.; Yuan, Y. *Radiat. Biol. Res. Commun.* **1988**, 23(4), 209–220.
29. Office for Research of Prevention and Treatment of Cancer; Data Collection of Malignant Neoplasm Mortality in China; Beijing, China, 1980.
30. Wang, Zuoyuan et al. *J. Natl. Cancer Inst.* **1990**, 82(6), 478–485.
31. Shimizu, Y.; Kato, H.; Schull, W. J.; Preston, D. L.; Fujita, S.; Pierce, D. A. *Comparison of Risk Coefficients for Site-Specific Cancer Mortality*

Based on the DS 86 and T65DR Shielded Kerma and Organ Doses; Life Span Study Report 11, Part .1: RERF TR 12-87, 1987.

RECEIVED for review August 7, 1992. ACCEPTED revised manuscript March 23, 1993.

Health and Mortality among Contractor Employees at U.S. Department of Energy Facilities

Shirley A. Fry[1], Donna L. Cragle[1], Douglas J. Crawford-Brown[2], Elizabeth A. Dupree[1], Edward L. Frome[3], Ethel S. Gilbert[4], Gerald R. Petersen[5,7], Carl M. Shy[2], William G. Tankersley[1], George L. Voelz[6], Phillip W. Wallace[1], Janice P. Watkins[1], James E. Watson, Jr.[2], and Laurie D. Wiggs[6]

[1]Oak Ridge Institute for Science and Education, Oak Ridge, TN 37831
[2]University of North Carolina, Chapel Hill, NC 27514
[3]Oak Ridge National Laboratory, Oak Ridge, TN 37831
[4]Pacific Northwest Laboratory, Richland, WA 99352
[5]Hanford Environmental Foundation, Richland, WA 99352
[6]Los Alamos National Laboratory, Los Alamos, NM 87545

Since 1978 follow-up studies of plant-specific and combined populations involving ~360,000 current and former employees of the U.S. Department of Energy (DOE) and predecessor agencies and their contractors at 40 or more sites nationwide have been conducted by DOE contract epidemiologists as part of the Health and Mortality Study of Atomic Workers. Among these populations, death rates to date for all causes of death combined and for most specific disease categories generally have been found to be similar to or lower than those in the U.S. population. No consistent pattern of increases in site-specific cancer mortality has been identified thus far across the populations studied. Although statistical associations have been demonstrated between certain cancer increases and employees' occupational radiation exposure, it is premature to draw conclusions about the contribution to their causation of occupa-

[7]Current address: Office of Health, U.S. Department of Energy, Washington, DC 20545

0065–2393/95/0243–0239$08.00/0

tional exposure to potentially hazardous agents given the gen-
erally low mortality and other study limitations. A summary
review is presented of completed and ongoing studies in this
series.

$\mathbf{M}$EDICAL MONITORING OF WORKERS employed by contractors operating facilities for the Manhattan Engineer District (MED) was initiated with the start-up of operations at individual facilities beginning
in 1942. These programs were aimed at protecting the health of active
workers against the short-term effects of exposure to the major toxicants present in the workplace. Ionizing radiation from external sources
or internally deposited radionuclides was the primary hazard of interest. However, workers at risk of exposure to certain chemical toxicants, including uranium and other heavy metals, were assigned to
specific medical monitoring or bioassay programs, or both. In the 1950s
greater awareness of the long-term health risks of exposure to sub-
acute levels of radiation and the increasing use of radiation and radioactive materials for industrial and medical purposes, and in related
research and development activities prompted interest by the Atomic
Energy Commission (AEC), the MED's successor agency, in conducting long-term follow-up studies to better protect the health of current and future workers overall and with respect to the delayed effects
of radiation. These and other rationales for and the importance of long-
term studies of populations occupationally exposed to radiation have
been discussed by Shore (1).

The purposes of this chapter are as follows: (1) to put the development of the long-term studies of AEC contractor employees prior
to 1978 into historical perspective; (2) to describe the scope and nature
of the studies of this population, which we and other epidemiology
groups have conducted since 1978 in the Department of Energy's
(DOE's) epidemiology program as part of its Health and Mortality Study
(HMS) as described by Thomas (2); and (3) to summarize the findings
of the studies completed to date and their implications for society.

Historical Perspective, 1960–1977

In the early 1960s a series of studies was conducted to evaluate the
feasibility of using plant personnel and other records as the basis for
long-term follow-up studies to monitor mortality among employees of
two contractors of the AEC. The first of these series of studies concerned small groups of uranium workers at the MED and AEC sites
at Mallinckrodt Chemical Works plants in St. Louis and Weldon

Springs, Missouri, and the Feed Materials Production Center operated for the AEC by National Lead of Ohio at Fernald Ohio. The studies were conducted for these AEC contractors by the University of Colorado, Boulder, and demonstrated the feasibility of using plant records for the stated purpose. The results were presented in a series of reports (3–7).

In 1964 the AEC initiated a 5-year pilot project that was more broadly based but with objectives similar to those of the earlier studies. This project was conducted by the University of Pittsburgh under contract from the AEC with T. Mancuso as the principal investigator. Selected for inclusion in this project were the MED–AEC facilities at Hanford, Washington, and Oak Ridge, Tennessee, and several uranium feed materials and conversion facilities that were the responsibility of the Oak Ridge Operations, including Harshaw Chemical Company (Cleveland, Ohio), Mallinckrodt Chemical Works, and the Feed Materials Production Center. Manhattan Engineer District contractor employees at the University of Chicago's Metallurgical Laboratory, and DuPont also were included in the feasibility study (Marks, S. AEC, unpublished data). Original plant records were located and identified at the Hanford and Oak Ridge sites and at off-site federal and other records repositories. Much of the data needed for retrospective epidemiologic studies was identified among the original hard copy employee records compiled for the MED and AEC by facility contractors. At Mancuso's request the AEC placed a moratorium on the disposition of such records, and thus preserved them for future epidemiologic studies.

During this period Mancuso established interfaces between the AEC and the Social Security Administration (SSA) for determination of vital status (alive vs. deceased) for individual workers and agreements with the Vital Records Offices of each of the 50 states for the retrieval of the corresponding death certificates as sources of cause of death information. The University established an office in Oak Ridge to direct the retrieval of plant personnel records and death certificates and the processing of relevant data into a machine-readable form. The data computerization task was performed under Mancuso's direction at the data processing facilities operated for the AEC in Oak Ridge, Tennessee, by Union Carbide Corporation–Nuclear Division.

This work established the apparent feasibility of using existing employee and other facility records as the basis for follow-up studies to monitor the health and mortality experience of workers employed at its contractor operated facilities and those formerly operated for the MED and to determine if any adverse effects observed were related to their employment at these facilities. It also confirmed the suitability of existing records for use in studies designed to estimate the upper bound of the cancer risk associated with occupational exposure to low levels of ionizing radiation.

In 1970, and on the basis of the results of the feasibility study, the AEC contracted with the University of Pittsburgh to have Mancuso initiate a long-term study of health and mortality among the populations identified in the pilot phase. The records used for the feasibility study and subsequent epidemiologic studies of these worker populations currently are retained for the DOE in accordance with regulations governing the DOE Systems of Records whereby they are protected under the Privacy Act (1974) (8).

As the work progressed it became evident that the original data did not meet the investigators' initial expectations of immediate usability for epidemiologic purposes and indicated that further editing, verification, and other processing of the data would be necessary to ensure their completeness, epidemiologic validity, and usability in analyses. A major effort then was directed toward preparing the Hanford worker population data for analysis. Analyses of longevity among Hanford workers were begun by B. Sanders, consultant statistician to the University of Pittsburgh team. Beginning in 1971 these analyses were documented along with other project-related activities in the investigators' annual progress reports to the AEC (9–21); the results subsequently were published in the scientific literature (22). During this period the investigators proposed extending the scope of the study to include workers at other selected contractor facilities. Specifically, employee populations from the following facilities were identified as priorities for inclusion in the overall study (Marks, S. AEC, unpublished data): Los Alamos Scientific (now National) Laboratory (LANL), Los Alamos, New Mexico; Rocky Flats Plant, Golden, Colorado; and Mound Laboratory, Miamisburg, Ohio.

In 1974 S. Milham, Washington State Department of Social and Health Services, reported finding an increased proportion of deaths due to cancer among Hanford workers on the basis of a broad proportional mortality analysis by occupation as recorded on death certificates filed in Washington State between 1950 and 1971 (23). Following this report, and on the basis of findings of a peer-review panel sponsored by the AEC, the Commission decided to terminate its contract with the University of Pittsburgh effective at the end of August 1977. In March 1975 Mancuso was notified informally of this decision. In January 1976 Mancuso received written notification of this decision (Marks, S. AEC, unpublished correspondence). The interval between the notification and the planned termination of the contract was designated as a transition period. During this 18-month period, the AEC prepared to transfer responsibilities for the continuation of work on the study to other contractors.

The actions of the AEC and its successor agency, the Energy Research and Development Administration (ERDA), were later investigated by the U.S. Comptroller General. The findings of the investigation, which generally upheld the ERDA's position, were reported to the U.S. Congress in 1979 (24). Initial analyses of mortality among Hanford workers alone were conducted by the University of Pittsburgh team headed by Mancuso, and the results were published in 1977 (25). The reported findings of increased risks for several different cancer types associated with the population's occupational radiation exposure were unexpected on the basis of contemporary knowledge.

At least 20 reviews and critiques of the Mancuso, Stewart, and Kneale Study (24), including those by Reissland (26) and Anderson (27), and reports of the reanalyses of the Hanford data set using established epidemiologic and biostatistical methods were generated in response to the 1977 publication. The results of the major reanalyses of the data set (28, 29) generally did not support the findings reported by Mancuso et al. Updated analyses by a team of investigators from the Hanford Environmental Health Foundation (HEHF) and Pacific Northwest Laboratory (PNL) of mortality among the Hanford worker population also failed to support the findings reported by Mancuso et al. of increasing risks for several types of cancer with increasing radiation dose, except for multiple myeloma, which continued to be associated statistically with occupational radiation dose in subsequent updated analyses (30–32).

Health and Mortality Study, 1977–1990

During the transition period between January 1976 and July 1977, responsibilities for follow-up studies of specific plant populations already selected for inclusion in the HMS were reassigned by the ERDA and its successor the DOE to its contractors Oak Ridge Associated Universities (ORAU), Oak Ridge, Tennessee, LANL, and the HEHF–PNL team, Richland, Washington. The ORAU was charged with developing studies of the worker populations at the production and research and development facilities in Oak Ridge, Tennessee [i.e., the Y-12 and Gaseous Diffusion (K-25) plants and Oak Ridge National Laboratory (ORNL, also known as X-10)]; the Gaseous Diffusion Plants in Paducah, Kentucky, and Portsmouth, Ohio; Feed Materials Production Center; and the MED/AEC sites of Mallinckrodt Chemical Works. Uranium processing, enrichment, and metal fabrication operations were common to several of these facilities.

In addition, ORAU was assigned responsibility for development of a master roster of contractor employees at the identified facilities, management of the DOE–SSA interface for the contractor epidemiology groups, and related data collection and management, including death certificate retrieval and storage. From 1979 ORAU worked in collaboration with the University of North Carolina—Chapel Hill (UNC—CH), School of Public Health, to provide additional academic guidance and assistance in all tasks, except those involving identifiable data. At that time ORAU also was charged by the DOE with developing a separate roster of contractor employees reported to have received whole-body doses of 50 mSv or more of external penetrating radiation in any year while employed from 1947 onward at any DOE or contractor facility nationwide (Lenhard, J. A., DOE Oak Ridge Operations, unpublished correspondence, May 22, 1978).

The LANL was assigned the DOE's National Plutonium Workers' Study and studies of the entire workforces of several sites at which workers were monitored for exposure to plutonium (i.e., LANL, Rocky Flats Plant, and Mound Laboratory). Plutonium workers identified among workers at ORNL, the entire Savannah River Plant workforce, which was studied collaboratively later by ORAU and LANL, and the Hanford population also were included in LANL's Plutonium Workers' Study. The HEHF–PNL team was assigned responsibility for continuing data collection and processing (HEHF) and data analysis (PNL and HEHF) for the total Hanford worker population.

The results of a survey by the Mitre Corporation, McLean, Virginia, to identify active and former contractor sites of the MED and its successor agencies were published in 1978 (33). This survey also included summary descriptions of records of epidemiologic interest that were available for facility-specific populations.

In 1979 the DOE accepted in principle ORAU's proposal entitled "The Comprehensive Epidemiology Study of Atomic Workers" to expand the AEC–DOE HMS to include all workers at all active and inactive DOE contractor sites. Seventy-six such sites, with an estimated total workforce of 600,000, were identified from the Mitre Corporation report as being eligible for eventual inclusion in the study.

The active and inactive facilities, whose present and former employee populations were included in the DOE Comprehensive Epidemiology Study of the "Atomic Workers" proposal, are identified in Figure 1, the so-called "big picture", although not all have come under active study. A description of the components of this figure is provided elsewhere (34). The capability of bringing all site-specific worker populations under active study was a function of the resources available for the task.

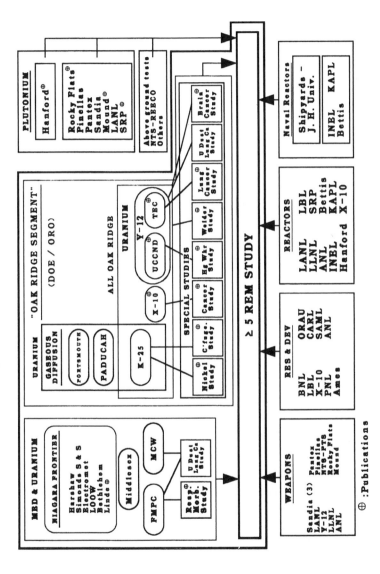

Figure 1.   The "big picture", representing the federal facilities identified for inclusion in the Comprehensive Epidemiology Study of Atomic Workers (HMS). Special studies of selected subgroups of populations assigned to ORAU–UNC are identified.

An estimated 360,000 present and former workers comprise the portion of the total workforce (600,000) that was the basis for studies by ORAU, LANL, and HEHF–PNL through 1991. Data are not necessarily complete for all the estimated 360,000 workers nor have all these individuals been included thus far in population-specific analyses (see the following sections). However, efforts have been made to follow up all race and gender groups in this population to determine their vital status in preparation for future analyses.

Studies Conducted between 1978 and 1990 as Part of the HMS

Scope. Since 1979 investigators from ORAU in collaboration with those from the Departments of Epidemiology and Environmental Sciences and Engineering of UNC—CH, have been engaged in all phases of data collection, verification, editing, processing, and management for plant-specific populations as assigned (*see* the preceding discussion). The DOE subsequently assigned responsibility to ORAU for studies of mortality among the workforces at six former MED uranium processing and refining facilities in the "Niagara Frontier" area, including Harshaw Chemical Company, Cleveland, Ohio, and for studies of disease incidence (morbidity) and mortality among the workers identified as having received 50 mSv or more in any calendar year while employed at any DOE (or predecessor) contractor facility nationwide (*see* the foregoing discussion). In addition, in 1982 and 1986, respectively, the DOE directed ORAU to compile data as a basis for studies of mortality among workers at the Savannah River Plant (SRP), the Lawrence Livermore National Laboratory (LLNL), and the Lawrence Berkeley Laboratory (LBL). The task was completed for the SRP population and has been ongoing for the LLNL population in collaboration with researchers at the LLNL. Initiation of data collection for the entire LBL population was postponed pending the availability of additional resources. The Oak Ridge Associated Universities' plant-specific study populations thus totaled approximately 260,000 individuals at 10 geographically separate sites, plus the almost 40 sites at which workers with 50 mSv or more in a calendar year were identified. Of these, approximately 138,000 (primarily white males) are included in populations defined in published and ongoing studies.

Approach. The principal purpose of the HMS was to evaluate the effects, on subsequent health and mortality, of occupational exposure to low levels of ionizing radiations from external and internally deposited radionuclides (e.g., uranium and plutonium) alone or in the presence of chemical toxicants in the workplace (e.g., uranium compounds and toxic metals) that may influence radiation-induced effects.

Early in ORAU's involvement in the HMS, a standardized study process (SSP) was developed to ensure consistency in the way in which ORAU investigators collected, compiled, and processed workers' data for inclusion in the HMS master roster (35). This approach was implemented to minimize the potential for introducing systematic bias into the data when working concurrently with data from multiple facilities. By this process, identifying, demographic, employment, work history, and personal monitoring data were retrieved systematically for all individuals ever employed at the facilities of interest irrespective of gender or race. Sources of these data were the employee records previously compiled by facility contractors for payroll, regulatory, or other nonepidemiological purposes. Data items contained in employee medical files were abstracted only as needed for specific studies of defined populations or subgroups. These data were entered into the computerized HMS database that contains one record per individual identifiable by an assigned unique numerical identifier (ID) across database files and facilities (if employed at more than one DOE contractor facility). Deidentification of the data maintained for individual employees in the computerized database files ensures their confidentiality and facilitates their use in statistical analyses.

Vital status information also was obtained systematically for all gender and race groups identified in the master roster. The SSA has been the primary source of information about deaths that occurred before 1979. Other sources include states' death indexes, the National Death Index (identifies deaths post-1978), states' departments of motor vehicles and drivers' license bureaus, the Office of Personnel Management (for federal employees), Pension Benefit Information, Inc., the Health Care Financing Agency (for persons aged 65 years and over), and the Veterans' Administration. Death certificates for persons identified as deceased by these or related institutions were retrieved under agreements of confidentiality from the vital records offices of the states of death. The underlying cause of death and all contributory cancer causes documented on verified death certificates were coded to the International Classification of Disease, adapted for use in the United States, Eighth Revision, by experienced nosologists trained by the National Center for Health Statistics. These data, identified by pseudoidentifiers, also are maintained in the HMS database. Mortality has been the end point of interest in the majority of studies conducted to date. Morbidity, as determined by telephone health surveys or clinical examinations, has been studied among fewer and more highly selected_study populations (36–38) because of the relative complexity and expense of such studies. However, greater emphasis on morbidity was planned for future studies of more defined populations.

Exposures of interest in the study populations have included radiation both from external sources and internally deposited radionuclides with uranium compounds (a primary interest of the ORAU–UNC investigators), being chemically as well as radiologically toxic, depending on the level of exposure, solubility, and specific activities of the compounds involved. Other nonradiological toxicants of interest in special studies included elemental mercury and metallic nickel (37, 39, 40). Nonoccupational exposures such as smoking were taken into account in only a few studies completed by any of the contractor epidemiology groups through 1990, but the influence of smoking was considered in a case-cohort study of lung cancer among Hanford workers (41). It also is being considered in a study in progress of lung cancer among workers exposed to uranium dust while employed at the Y-12, Fernald, and Mallinckrodt facilities. Data for workers identified as occupationally exposed to plutonium or polonium at Rocky Flats or Mound were analyzed as part of studies of mortality among these populations conducted by the LANL epidemiology group (42–44).

In compliance with the Privacy Act (1974) (8) and the regulations governing the release of personally sensitive and confidential data by the SSA and states' vital records offices, investigators' access to identifiable and other certified data for individual members of facility-specific populations being studied by ORAU–UNC has been restricted to those working at ORAU's Center for Epidemiologic Research and whose work requires access to these types of data, for example, for merging data retrieved for individuals from multiple sources.

Study Designs.   ORAU–UNC's overall study approach has called for initial hypothesis-generating analyses to compare the age- and sex-adjusted mortality rates in the worker populations, by facility, with those among the general (i.e., federal, state, or other regional population, as indicated) and other more appropriate comparison populations (e.g., workers at the same site). As occupational data were retrieved and prepared for analyses, hypothesis-testing studies were initiated among subgroups identified within or across plant-specific populations to evaluate relationships between exposures or jobs and diseases, particularly cancers, having statistical, radiobiological, or epidemiological significance. Industry or internal plant-specific populations were preferred comparisons in these analyses. Biostatistical methods to support these analyses were proposed initially at a DOE Statistical Symposium (45) and later developed in more detail (46–51). By this approach preliminary facility-specific mortality analyses could be performed to determine how the workers' mortality experience compared with that of the general population and to generate hypotheses that could be tested as more data became available. Vital status information for at least 90% of the study cohort and retrieval of over 90% of the death certificates for persons known to be dead were prerequisites for a preliminary SMR analysis to proceed. The LANL and HEHF–PNL epidemiology groups have employed generally similar approaches in their studies of other site-specific populations.

Cohort studies were conducted by the DOE contractor epidemiology groups to evaluate overall mortality among facility-specific populations at the Oak Ridge (Y-12 and ORNL) (52–55), Savannah River (56), Linde (57), Hanford (29–31), Rocky Flats (42), and Mound (43, 44) facilities. Studies of mortality among cohorts of workers at the Pantex Plant, Amarillo, Texas, and the Gaseous Diffusion Plant, Portsmouth, Ohio, were sponsored outside the HMS by the National Institute for Occupational Safety and Health (58, 59). Several of these studies used internal comparison groups, thereby minimizing the so-called "healthy worker effect" (60), to evaluate dose–response relationships (51). Cohort mortality studies are in progress for populations at all Oak Ridge facilities combined, LANL, Mallinckrodt Chemical Works, and Fernald, while updated mortality analyses are ongoing for previously defined populations at the Savannah River, ORNL, Y-12, and Hanford facilities.

Studies of mortality among subcohorts defined on the basis of job title or potential for exposure to nonradioactive agents of interest were conducted for workers employed in the centrifuge monitored for exposure to mercury (39) or phosgene (61) at the Y-12 Plant or nickel at the K-25 plant (40). Mortality through 1973 among welders at the facilities in Oak Ridge was evaluated and the results were reported (62); an updated analysis is in progress and includes deaths among the Oak Ridge welder subcohort through 1989.

Case-control study designs were used to evaluate specific cancers of interest found in greater numbers than expected. These include studies of deaths due to brain tumors among the Rocky Flats population (63), brain cancers among the Y-12 and ORNL populations (64), and lung cancer among workers employed at Y-12 (65) or Hanford (41). Case-control studies of cases of and deaths due to melanoma were conducted among employees at the LLNL and LANL facilities, respectively (66, 67).

Morbidity was evaluated among workers potentially exposed to elemental mercury at Y-12 (37) and among workers employed in the centrifuge process at the K-25 plant (38). To date, morbidity data have been collected for workers with 50 mSv or more in a calendar year at 20 of the 31 DOE contractor facilities at which such workers were identified.

As identified in the protocol proposed for the Comprehensive Epidemiology Study of Atomic Workers, an ultimate goal of the study has been to conduct combined population analyses where feasible and epidemiologically appropriate in order to increase the power of the analysis and thereby the strength of the results and the precision of the estimates of risks for radiation-induced cancers derived from them. Progress has been made in this direction with the completion of mortality analyses for the combined population of white males employed at the Hanford, ORNL, and Rocky Flats facilities (68). Data from already published facility-specific studies of these three populations also are included in an ongoing study, sponsored by the International Agency for Research on Cancer (IARC), of mortality among nuclear industry workers in the United States, Canada, and United Kingdom. Other studies completed or in progress involving workers at multiple DOE contractor facilities include cohort mortality analyses of workers employed between 1943 and 1947 at any of the Oak Ridge facilities (69) and, as previously mentioned, persons in the 50-mSv/year or more cohort and the case-control study of lung cancer deaths among workers at three uranium processing facilities. The publications referenced give the methodologic details of specific studies.

Study Findings. Some of the characteristics of the contractor facility-specific populations that have been studied to date are displayed in Tables I–III, grouped according to the primary exposure of interest. A summary of the mortality outcomes of interest for which statistically significant increases or deficits in the standardized mor-

Table I. DOE Worker Populations Studied

Facility (Years of Operation)	Total Workforce	Study Population	Follow-Up (Average Years)	Mortality (%)
Hanford (1943–1978)	44,100	44,100	23.0	20.0
ORNL (1943–1972)	17,500	8,318	26.0	18.3
Savannah River (1952–1975)	18,000	9,860	22.0	11.1
Rocky Flats[a] (1952–1979)	9,500	5,413	14.5	7.6
Pantex (1951–1978)	5,500	3,564	14.6	7.5
Mound Lab[a] (1943–1979)	6,880	4,182	18.8	14.2

[a]External ± internal exposures.

Table II. DOE Worker Populations Studied: Uranium Dust Exposures

Facility (Years of Operation)	Total Workforce	Study Population	Follow-Up (Average Years)	Mortality (%)
Oak Ridge Y-12 (1943–1947)	48,000	18,800	27.0 (minimum)	28.6
Oak Ridge Y-12 (1947–1972)	16,500	6,477	20.6	13.0
Niagara Frontier (Linde; 1943–1949)	3,000	995	30.0 (minimum)	43.0

Table III. DOE Worker Populations Studied: Uranium Hexafluoride

Facility (Years of Operation)	Total Workforce	Study Population	Follow-Up (Average Years)	Mortality (%)
Oak Ridge K-25 (1943–)	45,000	←———— Work in progress ————→		
Paducah GDP (1952–)	6,000	←———— Work in progress ————→		
Portsmouth GDP (1954–1982)	7,900	5,773	18.7	8.4

tality ratios have been found is presented in Table IV. These findings simply reflect the populations' mortality experience for all cancers combined and site-specific cancers relative to the general population, taking age and gender into account.

Analyses of all or site-specific cancer mortality, taking occupational radiation dose into account, have been conducted since 1978 for several facility-specific and combined facility populations (*30–32, 42–44, 53–55, 57, 64, 65, 68*). Statistically significant positive associations between mortality due to certain cancers and occupational radiation dose were found in two instances. A positive association between multiple myeloma and radiation dose was reported in updated studies of the Hanford population (*30–32*). However, in the third study of this series, the association of multiple myeloma and radiation dose reached significance only when deaths were included for which multiple myeloma was shown on the death certificates but not as the underlying cause, when deaths that occurred in the 2 years after the established cutoff date for follow-up, or when a 2-year latency period was used.

Analyses of mortality among a cohort of white males employed at the ORNL showed overall and some site- and disease-specific mortality to be substantially less than the general population, but a non-statistically significant elevated SMR for all leukemias was reported with follow-up through 1977 (*54*). This parameter attained statistical significance with further follow-up through 1984 (*55*). However, neither study showed a positive correlation for leukemia with radiation

Table IV. Major End Points of Interest for Which the Ratio
of the Number of Observed to Expected Cases Was Statistically
Significant (Significance Level Established a Priori)

Cause of Death	Facility[a]
I. Increases	
All causes	Linde, O.R. WW II[b]
Cardiovascular disease	Linde
Respiratory disease	Linde, O.R. WW II
Lung	Y-12, O.R. WW II
Cancer	(Rocky Flats; Mound)
Larynx	Linde
Leukemias	ORNL (Savannah River)
Brain	(Y-12)
Melanoma[c]	Livermore
Accidents	(Hanford); O.R. WW II
II. Deficits	
All causes	Y-12; Savannah River; Rocky Flats; ORNL; Mound
All cancers	Savannah River; ORNL; Rocky Flats
Lung	ORNL, Rocky Flats
Digestive system	O.R. WW II
Cardiovascular disease	Savannah River; ORNL; Rocky Flats; Mound; Portsmouth
Respiratory disease	Savannah River; ORNL; Rocky Flats; Portsmouth
Digestive disease	Savannah River
Accidents	ORNL; Rocky Flats; Mound; Portsmouth

[a]Facility: Indicates statistically significant SMR among a subgroup of the study population at the facility named.
[b]This represents the combined population of workers employed only between 1943 and 1947 at the Y-12, K-25, and ORNL facilities.
[c]This corresponds to the incidence of melanoma only.

dose. However, with follow-up through 1984, a significant positive association was observed between external radiation, assuming a 20-year lag, and all cause mortality that was primarily due to an association with all cancer mortality. One death due to bone cancer was reported among a group of 26 early plutonium workers (70), but a causal relationship between this individual's malignancy and exposure to plutonium remains equivocal given the small number.

The results of the combined population mortality analyses completed to date are generally consistent with those obtained in the mortality analyses conducted for the plant-specific populations concerned (51, 68). Among the case-control studies of specific cancers completed to date, only the study of lung cancer among workers employed at Y-12 between 1943 and 1947 showed an increased risk that was statistically significantly associated with the occupational exposure to radiation, with cumulative radiation dose to the lung estimated based on occupational exposure to uranium dust. The association was significant only for workers who were more than 45 years old when hired at the facility (65).

Discussion

A review of the results of the long-term studies of mortality that have been completed to date for 10 plant-specific populations of the DOE (and predecessor agencies) contractor employees shows that, with the exception of two cohorts of workers employed during World War II [i.e., Linde and Oak Ridge WW II (57, 69)] overall mortality due to all and specific causes of death, including cancers of all and most specific sites, is generally similar to or substantially less than expected among comparable groups in the U.S. population. The mortality experiences of the Linde and Oak Ridge WW II cohorts possibly reflect relaxed medical standards for civilian employment during the war years. Although statistically significant increases in mortality due to specific cancers were observed among some of the other plant-specific cohorts, no consistent pattern of increased cancer mortality was found across facilities. Where statistically significant associations were observed between risks of mortality caused by all cancers combined or type-specific cancers, the associations generally were weak and inconsistent across populations.

In evaluating the reported results of these studies, we need to consider the overall strengths and limitations that may contribute to uncertainties among the findings to date. Strengths include the fact that the populations studied are composed of humans, and the study results relate directly to risks to human health. Because they were worker populations, we can assume, with the possible exception of Linde and Oak Ridge WW II workers, that the study participants were healthy when hired and that the mortality deficits observed may be attributed to the healthy worker effect. These populations were at risk of protracted exposure to radiation, which is in contrast to the high-dose, high-dose-rate exposures sustained by populations such as the atomic-bomb survivors and patient groups on whose mortality experience current radiation risk estimates are largely based.

Another strength of these studies is that records were available that included some individual quantitative measures of radiation dose or level of exposure for individual workers. This situation was not the case for some populations studied because of their exposure to radiation or other hazardous agents on or off the job; for these populations, only rough surrogates, (e.g., distance from source of radiation, job title, and location) are available to indicate potential exposure (71). Also, all populations for which mortality analyses have been conducted possess more than 90% vital status follow-up and 91–99% retrieval of verified death certificates. Among the limitations of the individual studies is the heterogenicity of the populations with respect to socioeconomic status and geographic location. To date there has been only limited consideration of the influence on health outcomes of known risk factors such as smoking (41), and undoubtedly, several risk factors are unknown. The results of these and future studies of these populations are generalizable to males, specifically white males, but they are not necessarily generalizable to other age and gender groups in the general population. However, for the major end points of interest with respect to radiogenic cancers, we would not anticipate effects of radiation among the populations of female workers to be substantially different from their male counterparts.

The estimated total workforce of 600,000 ever employed at the federal contractor sites of interest is one of the largest populations of radiation workers available and enumerated for study. However, the studies conducted to date have of necessity involved plant-specific populations that are relatively small. Studies of these individual worker populations are limited statistically in their ability to detect small risks to occupational radiation exposures. The power of the studies also is limited by the relatively low total population dose, even when populations are combined. An additional recognized limitation is that to date the majority of studies were based on death certificate data which in themselves are subject to uncertainty. However, the uncertainties in these data are nondifferentially distributed among the study populations and their comparison populations and, more importantly, among exposure groups. Some of these limitations may be overcome or reduced by inclusion of additional populations in combined analyses and the use of incidence data to the greatest possible extent.

Weighing all the available data to date, including those from studies that have been interpreted as indicating influence by factors not considered by other investigators (72, 73), it can be concluded that, overall, the DOE worker population exhibits a healthy worker effect relative to the general population. The completed studies provide little evidence that the workers' exposure to radiation on the job has significantly increased their risk of dying from cancer in a measurable quantity. These studies do not appear to present a strong case in favor of more stringent occupational radiation protection standards, although they currently may lack the power to detect a risk that could be judged unacceptable by today's standards.

Finally, even though the follow-up of mortality now extends more than 40 years for some populations, the mortality is generally less than 50% in most populations and less than 20% in several others. Thus, we recommend that further follow-up be undertaken in order to obtain health data on a majority of these workers through their lifespans, particularly with regard to the study of cancer rates that increase rapidly at older ages. The decision to pursue further studies, including any extended combined population studies, rests with the NIOSH, the agency now responsible for the management of the DOE's analytic epidemiologic studies.

Acknowledgments

This chapter reports work undertaken as part of the Health and Mortality Study on the U.S. DOE workers being conducted by the Oak Ridge Institute for Science and Education, in collaboration with the UNC, under Contract No. DE–ACO5–76OR00033 between the U.S. DOE and ORAU and in accordance with the Memorandum of Understanding between the DOE and the Department of Health and Human Services, December 1990.

This work reflects the efforts and support of many people in addition to us, specifically other members of the epidemiology groups involved, the responsible DOE Headquarters and Operations Offices staff, contractor management, labor and technical support staffs at the facilities studied, and the scientific advisors to the epidemiology groups, to whom we are sincerely grateful. Marta Rivera's assistance in the preparation of the manuscript also is greatly appreciated.

We also appreciate the vital statistics offices of the individual states as the sources of death record data and their technical support of this research. We are solely responsible for the data analyses and interpretation of the results.

References

1. Shore, R. E. *Health Phys.* **1990**, *59*(1), 63–68.
2. Chapter 4 in this book.
3. *Feasibility Study Report*; University of Colorado Medical Center: Denver, CO, 1962.
4. Bell, R. F.; Gilliland, J. C.; Bell, J. A.; Johnson, D. M. *A Report of an Epidemiological Feasibility Study at Three Uranium Feed Material Sites*; University of Colorado Medical Center: Denver, CO, 1965, Vol. I.
5. Bell, R. F.; Gilliland, J. C.; Bell, J. A.; Johnson, D. M. Figures and tables for *A Report of an Epidemiological Study at Three Uranium Feed Material Sites*; University of Colorado Medical Center: Denver, CO, 1965, Vol. II.
6. Bell, R. F.; Gilliland, J. C.; Bell, J. A.; Johnson, D. M.; Evans, E. E. *Epidemiologic Feasibility Report*; University of Colorado Medical Center: Denver, CO, 1966.
7. Quigley, J. A. *Epidemiological Study of Uranium Workers—Feasibility*; National Lead of Ohio: Cincinnati, OH, 1967.
8. *Fed. Regist.* **1982**, *47*, 14270–14285.
9. Mancuso, T. F.; Sanders, B. S.; Brodsky, A. *Feasibility Study of the Correlation of Lifetime Health and Mortality Experience of AEC and AEC*

Contractor Employees with Occupational Radiation Exposure; AEC Contract No. AT(30-1)-3394; University of Pittsburgh: Pittsburgh, PA, 1965.
10. Mancuso, T. F.; Sanders, B. S.; Brodsky, A. *Feasibility Study of the Correlation of Lifetime Health and Mortality Experience of AEC and AEC Contractor Employees with Occupational Radiation Exposure*; AEC Contract No. AT(30-1)-3394; University of Pittsburgh: Pittsburgh, PA, 1966.
11. Mancuso, T. F.; Sanders, B. S.; Brodsky, A. *Feasibility Study of the Correlation of Lifetime Health and Mortality Experience of AEC and AEC Contractor Employees with Occupational Radiation Exposure*; AEC Contract No. AT(30-1)-3394; University of Pittsburgh: Pittsburgh, PA, 1967.
12. Mancuso, T. F.; Sanders, B. S.; Brodsky, A. *Feasibility Study of the Correlation of Lifetime Health and Mortality Experience of AEC and AEC Contractor Employees with Occupational Radiation Exposure*; AEC Contract No. AT(30-1)-3394; University of Pittsburgh: Pittsburgh, PA, 1968.
13. Mancuso, T. F.; Sanders, B. S.; Brodsky, A. *Feasibility Study of the Correlation of Lifetime Health and Mortality Experience of AEC and AEC Contractor Employees with Occupational Radiation Exposure*; AEC Contract No. AT(30-1)-3394; University of Pittsburgh: Pittsburgh, PA, 1969.
14. Mancuso, T. F.; Sanders, B. S.; Brodsky, A. *Feasibility Study of the Correlation of Lifetime Health and Mortality Experience of AEC and AEC Contractor Employees with Occupational Radiation Exposure*; AEC Contract No. AT(30-1)-3394; University of Pittsburgh: Pittsburgh, PA, 1970.
15. Mancuso, T. F.; Sanders, B. S.; Brodsky, A. *Feasibility Study of the Correlation of Lifetime Health and Mortality Experience of AEC and AEC Contractor Employees with Occupational Radiation Exposure;* AEC Contract No. AT(30-1)-3394; University of Pittsburgh: Pittsburgh, PA, 1970.
16. Mancuso, T. F.; Sanders, B. S.; Brodsky, A. *Feasibility Study of the Correlation of Lifetime Health and Mortality Experience of AEC and AEC Contractor Employees with Occupational Radiation Exposure;* AEC Contract No. CH AT(11-1)-3428; University of Pittsburgh: Pittsburgh, PA, 1972.
17. Mancuso, T. F.; Sanders, B. S.; Brodsky, A. *Feasibility Study of the Correlation of Lifetime Health and Mortality Experience of AEC and AEC Contractor Employees with Occupational Radiation Exposure;* AEC Contract No. AT(11-1)-3428; University of Pittsburgh: Pittsburgh, PA, 1973.
18. Mancuso, T. F.; Sanders, B. S.; Brodsky, A. *Feasibility Study of the Correlation of Lifetime Health and Mortality Experience of AEC and AEC Contractor Employees with Occupational Radiation Exposure;* AEC Contract No. AT(11-1)-3428; University of Pittsburgh: Pittsburgh, PA, 1974.
19. Mancuso, T. F.; Sanders, B. S.; Brodsky, A. *Feasibility Study of the Correlation of Lifetime Health and Mortality Experience of AEC and AEC Contractor Employees with Occupational Radiation Exposure;* AEC Contract No. E(11-1)-3428; University of Pittsburgh: Pittsburgh, PA, 1975.
20. Mancuso, T. F.; Sanders, B. S.; Brodsky, A. *Feasibility Study of the Correlation of Lifetime Health and Mortality Experience of AEC and AEC Contractor Employees with Occupational Radiation Exposure;* AEC Contract No. E(11-1)-3428; University of Pittsburgh: Pittsburgh, PA, 1976.
21. Mancuso, T. F.; Sanders, B. S.; Brodsky, A. *Feasibility Study of the Correlation of Lifetime Health and Mortality Experience of AEC and AEC Contractor Employees with Occupational Radiation Exposure*; AEC Contract No. EY-76-S-02-3428; University of Pittsburgh: Pittsburgh, PA, 1977.
22. Sanders, B. S. *Health Phys.* **1978**, *34*, 521–528.

23. Milham, S., Jr. *Occupational Mortality in Washington State, 1950–1971;* Contract CDC-00-74-26; State of Washington: Olympia, WA, 1975.
24. Comptroller General. *Report to the Congress of the United States: Review of the Department of Energy's Controversial Termination of a Research Contract*, EMD-79-21; U.S. General Accounting Office: Washington, DC, 1979.
25. Mancuso, T. F.; Stewart, A.; Kneale, G. *Health Phys.* **1977,** *33,* 369–384.
26. Reissland, J. A. Report No. NRPB-R79, 1978; National Radiological Protection Board: Harwell, Didcot, United Kingdom.
27. Anderson, T. W. *Health Phys.* **1978,** *34,* 743–750.
28. Hutchison, G. B.; MacMahon, B.; Jablon, S.; Land, C. E. *Health Phys.* **1979,** *37,* 207–220.
29. Gilbert, E. S.; Marks, S. *Radiat. Res.* **1979,** *79,* 122–148.
30. Gilbert, E. S.; Marks, S. *Radiat. Res.* **1980,** *83,* 740–741.
31. Tolley, H. D.; Marks, S.; Buchanan, J. A.; Gilbert, E. S. *Radiat. Res.* **1983,** *95,* 211–213.
32. Gilbert, E. S.; Petersen, G. R.; Buchanan, J. A. *Health Phys.* **1989,** *56,* 11–25.
33. Harlow, M.; Anderson, C.; Rader, R. Report M78-52, 1978; Mitre Corporation: McLean, VA, 1978.
34. Lushbaugh, C. C. *The Development and Present State of the DOE Health and Mortality Studies;* From DOE Radiation Epidemiology Contractors' Workshop: Program and Working Papers. U.S. Department of Energy: Washington, DC, 1982; pp 1–36.
35. Lushbaugh, C. C.; Fry, S. A.; Shy, C. M.; Frome, E. L. *Proc. 16th Midyear Topic. Symp., Health Phys.* **1983,** 105–114.
36. Fry, S. A.; Rudnick, S. A.; Hollis-Hudson, D. R.; Robie, D. M.; Lushbaugh, C. C.; Shy, C. M. *Proc. 3rd Int. Symp. Rad. Protect.* **1982,** *1,* 346–351.
37. Fine, L. *Health Evaluation of Y-12 Workers Formerly Exposed to Mercury;* Final Report; University of Michigan: Ann Arbor, MI, 1987.
38. Cragle, D. L.; Wells, S. M.; Tankersley, W. G. *Appl. Occup. Environ. Hyg.* **1993,** *7,* 826–834.
39. Cragle, D. L.; Hollis, D. R.; Qualters, J. R.; Tankersley, W. G.; Fry, S. A. *J. Occup. Med.* **1984,** *26*(11), 817–821.
40. Cragle, D. L.; Hollis, D. R.; Shy, C. M.; Newport, T. H. In *Nickel in the Human Environment*, Lyon, France, 1983; Sunderman, F. W., Ed.; International Agency for Research on Cancer, Scientific Publication No. 53; IARC: Lyon, France, 1984; pp 57–63.
41. Petersen, G. R.; Gilbert, E. S.; Buchanan, J. A. et al. *Health Phys.* **1990,** *58*(1), 3–11.
42. Wilkinson, G. S.; Tietjen, G. L.; Wiggs, L. D.; Galke, W. A.; Acquavella, J. F.; Reyes, M.; Voelz, G. L.; Waxweiler, R. J., *Am. J. Epidemiol.* **1987,** *125,* 231–250.
43. Wiggs, L. D.; Cox-Devore, C. A.; Voelz, G. L. *Health Phys.* **1991,** *61*(1), 71–76.
44. Wiggs, L. D.; Cox-Devore, C. A.; Wilkinson, G. S.; Reyes, M. *J. Occup. Med.* **1991,** *33*(5), 632–637.
45. Frome, E. L.; Hudson, D. R. *Proc. 1980 DOE Statist. Symp.* **1980,** 206–218.
46. Frome, E. L. *Am. Statistician* **1981,** *35,* 262–263.

47. Frome, E. L. *Biometrics.* **1983,** *39,* 665–674.
48. Frome, E. L.; Checkoway, H. *Am. J. Epidemiol.* **1985,** *121,* 309–323.
49. Gilbert, E. S.; Buchanan, J. A. *J. Occup. Med.* **1984,** *26*(11), 822–828.
50. McLain, R. W.; Frome, E. L. *Proc. SAS Users Group Int. Conf.,* SUGI 11, Statistical Analysis Systems: Cary, NC, 1986; pp 874–876.
51. Gilbert, E. S. *Occupational Medicine: State of the Art Reviews;* Hanley and Belfus, Inc.: Philadelphia, PA, 1991; Vol. 6(4), pp 665–680.
52. Polednak, A. P.; Frome, E. L. *J. Occup. Med.* **1981,** *23,* 169–178.
53. Checkoway, H.; Pearce, N.; Crawford-Brown, D. J.; Cragle, D. L. *Am. J. Epidemiol.* **1988,** *127,* 255–266.
54. Checkoway, H.; Matthew, R. M.; Shy, C. M.; Watson, J. E.; Tankersley, W. G.; Wold, S. W.; Smith, J. C.; Fry, S. A. *Br. J. Ind. Med.* **1985,** *42,* 525–533.
55. Wing, S.; Shy, C. M.; Wood, J. L.; Wolf, S. W.; Cragle, D. L.; Frome, E. L., *J. Am. Med. Assoc.* **1991,** *265*(11), 1397–1402.
56. Cragle, D. L.; McLain, R. W.; Qualters, J. R.; Tankersley, W. G.; Fry, S. A. *Am. J. Ind. Med.* **1988,** *14,* 379–401.
57. Dupree, E. A.; Cragle, D. L.; McLain, R. W.; Crawford-Brown, D. J.; Teta, M. J. *Scand. J. Work, Environ. Health* **1987,** *13,* 100–107.
58. Acquavella, J. F.; Wiggs, L. D.; Waxweiler, R. J.; MacDonnell, D. G.; Tietjen, G. L.; Wilkinson, G. S. *Health Phys.* **1985,** *48*(6), 735–746.
59. Brown, D. P.; Bloom, T. *Mortality among Uranium Enrichment Workers;* National Institute for Occupational Safety and Health, Centers for Disease Control, U.S. Department of Health and Human Services: Cincinnati, OH, 1987.
60. McMichael, A. J. *J. Occup. Med.* **1975,** *18*(3), 168–186.
61. Polednak, A. P.; Hollis, D. R. *Toxicol. Ind. Health* **1985,** *1,* 137–147.
62. Polednak, A. P. *Arch. Environ. Health* **1981,** 235–242.
63. Reyes, M.; Wilkinson, G. S.; Tietjen, G.; Voelz, G. L.; Acquavella, J. F.; Bistline, R. *J. Occup. Med.* **1984,** *26*(10), 721–724.
64. Carpenter, A. V.; Flanders, D. W.; Frome, E. L.; Cole, P.; Fry, S. A. *Am. J. Public Health* **1987,** *77,* 1180–1182.
65. Cookfair, D. L.; Beck, W. L.; Shy, C. M.; Lushbaugh, C. C.; Sowder, C. L. *Proc. 16th Midyear Topic. Symp. Health Phys. Soc.* **1983,** 398–406.
66. Reynolds, P.; Austin, D. F. *West. J. Med.* **1985,** *142*(2), 214–221.
67. Acquavella, J. F.; Wilkinson, G. S.; Tietjen, G. L.; Key, C. R.; Stebbings, J. H.; Voelz, G. L. *Health Phys.* **1983,** *45*(3), 587–592.
68. Gilbert, E. S.; Fry, S. A.; Wiggs, L. D.; Voelz, G. L.; Cragle, D. L.; Petersen, G. R. *Radiat. Res.* **1989,** *120,* 19–35.
69. Frome, E. L.; Cragle, D. L.; McLain, R. W. *Radiat. Res.* **1990,** *123,* 138–152.
70. Voelz, G. L.; Lawrence, J. N. P. *Health Phys.* **1991,** *61*(2), 181–190.
71. Health Effects of Exposure to Low Levels of Ionizing Radiation; BEIR V; National Academy of Sciences: Washington, DC, 1980.
72. Kneale, G. W.; Mancuso, T. F.; Stewart, A. M. *Br. J. Ind. Med.* **1984,** *41,* 9–14.
73. Kneale, G. W.; Mancuso, T. F.; Stewart, A. M. *Br. J. Ind. Med.* **1981,** *38,* 156–166.

Received for review November 23, 1992. Accepted revised manuscript July 9, 1993.

Does Nuclear Power Have a Future?

John F. Ahearne

Sigma Xi, The Scientific Research Society, P.O. Box 13975, Research Triangle Park, NC 27709

Nuclear power for electricity generation is at a standstill in the United States. The United States has more nuclear power capacity than France and Japan combined and receives about 20% of its electricity from nuclear power. But no new plants have been ordered since 1978 and utilities are shutting down older plants before the end of their lifetimes. Plant orders stopped because electricity demand stopped growing at a high rate, nuclear plants were taking 12 years to build, were costing over $4 billion, and, once built, were not running very well. Safety and nuclear waste concerns added to these problems by generating public opposition to nuclear power. This chapter addresses whether new designs that are proposed to meet these problems are likely to succeed.

THE QUESTION, "DOES NUCLEAR POWER HAVE A FUTURE?" is tinged with pessimism. That there will be a future for nuclear power is implicitly doubted. However, this is not a uniformly held assumption. It certainly is not a uniformly desired condition—especially for the nuclear industry, which also argues that it is not the view of the American public, at least as transmitted by the U.S. Council on Energy Awareness (USCEA). The USCEA polling results indicate that the U.S. public believes nuclear energy should play a role in meeting America's future energy needs. In the spring of 1992, 35% of U.S. adults re-

0065–2393/95/0243–0259$08.54/0

sponded that this role should be very important and another 38% thought nuclear energy would be somewhat important. However, a more significant poll asks Americans what are the most important problems facing the United States today. Public concerns about energy peaked at 6% just before the Gulf War. Since then it has disappeared from the polls and did not show up in the fall of 1991. Consequently, statements about the importance of any energy source must take into account that the public is being asked about an area they don't see as a problem.

Nuclear plants had been described by their early supporters as being cheap producers of electricity over a long lifetime, so that the average annual cost of electricity would be low, although the initial capital investment would be high. Although the latter has certainly been borne out by experience, the former is being questioned. When the operators of Yankee Rowe in Massachusetts, a small, 180-megawatt electric (MWe) station that went into operation in 1961, announced they would close the plant rather than meet the Nuclear Regulatory Commission (NRC) requirements, many in the nuclear industry expressed concern. This plant was the first to begin the NRC procedure to extend its license beyond the original 40-year term. In February 1992 an arrangement was announced that would lead to shutting down San Onofre-1 in California by 1993, after 25 years of operation. Industry proponents urge development of new designs, to meet many of the questions raised by the public, investors, and utilities.

Before addressing new developments, some background is necessary. First, we must assume that nuclear fission will be used to generate substantial amounts of electricity in the future; that is, nuclear power will be important. If we do not make this assumption, it does not make much sense to discuss how that electricity might be generated.

Two questions then will be addressed:

1. Will there be—are there—qualitatively different new designs and concepts for nuclear power plants?
2. Will new plants, perhaps using new designs, be built in the United States in the foreseeable future?

Future Use

Nuclear power may be important for more than electricity generation. Other uses, or potential uses, for nuclear power include space propulsion and district or process heating.

Use of nuclear power for space propulsion is receiving renewed examination, as interest in missions to the moon and beyond is growing. This use is not for the small radioisotope thermoelectric generators (RTGs), the thermionic power sources that have been used for decades, nor is it for the in-space power plants used by the former Soviet Union for some of their satellites. Rather, these designs are Nuclear Engine for Rocket Vehicle Applications (NERVA) and post-NERVA designs that provide very high specific impulse rockets. Although technologically interesting these are unlikely to support a major part of the nuclear industry.

Use of nuclear power for heat generation has not yet proven to be economic on a wide scale. Cogeneration, production of electricity and process steam, is being done today. In Canada the four units of the Bruce A nuclear station in Ontario produce electricity and also process heat and steam for the nearby heavy water plants. In Czechoslovakia a four-unit station generates electricity and produces low-temperature heat (70–150°C) for heating, industrial, and agricultural uses. In West Kazachstan a breeder reactor is used to provide electricity and high-temperature steam to a desalination plant, and both the Russians and the Chinese are working on reactors solely for heating (1). Russia recently announced that it will begin construction on two 500-megawatt thermal (MWt) heat reactors.

China is reported to be developing two types of reactors to provide heat for district grids: an atmospheric-pressure, swimming pool reactor for up to 120 MWt and a low-pressure reactor for up to 500 MWt (2). The one recent U.S. effort, Midland, in Michigan, ended up, after many years of construction problems, being converted from a nuclear plant into a natural gas plant.

However, nuclear power plants are primarily used to generate electricity and represent a significant portion of world electricity generating capacity. In 1988 nuclear power represented 12.2% of world generating capacity and accounted for 17% of the electricity generated (3).

Although nuclear power is at a standstill in the United States, it remains the world leader in the number of plants and total generating capacity, as seen in Tables I and II.

Nuclear power is extremely important in several countries, although in no country would a source producing a quarter or more of the electricity be unimportant. As is widely known, France is the leader in the use of nuclear-generated electricity and exports nuclear-generated electricity to several of its neighbors. Table III indicates that nuclear power also is a major source of energy in at least 12 other countries.

**Table I. Operating Nuclear Units as
of December 31, 1990**

Rank	Country	Capacity (GWe)
1	United States	100.6
2	France	55.8
3	Former Soviet Union[a]	34.7
4	Japan	30.9
5	Germany	24.4
6	Canada	14.0
7	United Kingdom	11.5

NOTE: No other country had at least 10 GWe.
[a]Data available only for the combined total of all the republics comprising the former Soviet Union.
SOURCE: Data are from reference 4.

**Table II. Operating Nuclear Units as
of December 31, 1990**

Rank	Country	Number of Units
1	United States	112
2	France	56
3	Former Soviet Union	45
4	Japan	41
5	United Kingdom	37
6	Germany	26
7	Canada	20
8	Sweden	12

NOTE: No other country had at least 10 units.
SOURCE: Data are from reference 4.

The United States has many energy resources and a large distribution system in place to transport oil, gas, coal, and electricity around the country. We have a national electricity grid, which can transfer power within and among pools. Our natural gas pipeline system enables gas to be used nationwide, although U.S. gas is produced in only a few states and also is imported from Canada. Our rail and road network allows coal, gasoline, and liquid gas to be transported throughout the nation.

The U.S. advantages can be seen by comparison with Russia. Approximately 80% of Russia's energy use is in European Russia, but 80% of the fossil fuel reserves are in Asian Russia. The rail and road system is not good. However, the pipelines they already have in place, if laid end to end, would reach halfway to the moon. An energy group noted that 40% of Russia is hard to reach by road or rail. This group is pushing for more nuclear plants to be built in the inaccessible regions to compensate for the difficulty of transporting fossil fuels.

Although nuclear power provided only about one-fifth of the electricity for the United States as a whole, nuclear power represented more than 50% of the electrical generation for five states in 1989 and more than 25% for nine others.

Of course nuclear power does have one major purpose—to generate electricity. Although societies are becoming increasingly electrified, electricity generation remains behind other uses of energy in all countries. Table IV shows the contribution of nuclear power to the

Table III. Percent Electricity from Nuclear Power
in 1990

Rank	Country	% Nuclear
1	France	74.5
2	Belgium	60.1
3	Hungary	51.4
4	South Korea	49.1
5	Sweden	45.9
6	Switzerland	42.6
7	Spain	35.9
8	Bulgaria	35.7
9	Taiwan	35.2
10	Finland	35.0
11	Germany	33.1
12	Czechoslovakia	28.4
13	Japan	27.1
14	United States	20.6
15	Argentina	19.8
16	United Kingdom	19.7
17	Canada	14.8
18	Former Soviet Union	12.2

NOTE: Of the other ten countries with nuclear power plants, none obtained at least 6% of their electricity from nuclear power.
SOURCE: Data are from reference 4.

Table IV. Nuclear Contribution to Electricity and Total Primary Energy
1990

Rank	Country	Electricity % Nuclear	Total Primary Energy % Nuclear
1	Sweden	45.9	31.2
2	France	74.5	29.8
3	Switzerland	42.6	20.6
4	Finland	35.0	19.0
5	Belgium	60.1	18.9[a]
6	South Korea	49.1	14.4
7	Taiwan	35.2	13.1
8	Spain	35.9	12.7
9	Japan	27.1	11.2

[a]Includes Luxembourg.
SOURCE: Data are from references 4 and 5.

country's total primary energy consumption for those countries for which nuclear power contributes at least 10% of total primary energy. Perhaps ironically, Sweden leads. There are eight other countries for which nuclear power contributes more than 10% of total primary energy consumption.

Obviously, nuclear power does have a future in supplying electricity, but the industry also has reason for concern, as indicated in Table V. Nuclear power no longer seems to be an option for new electricity generation.

New Designs

Many followers of nuclear power are calling for new designs. These calls are coming from both supporters and critics of nuclear power. Following are two examples.

In 1989, Jan Beyea, senior staff scientist of the Audubon Society, in testimony to Congress, said (6):

> . . .we are reluctant to put all our long-range, energy-supply eggs in the solar basket, so some level of research into the technological potential of "inherently safe" nuclear power is warranted . . . the major goal of any second-generation nuclear program must be restoration of public confidence . . . it is important that Congress lay out a tough design standard for second-generation reactors.. . . If engineers and scientists are held to a tough technical standard, they will rise to the challenge, meeting the goal, if it is at all possible to do so.

Alvin Weinberg and Charles Forsberg, from Oak Ridge, state (7):

> The accident at Chernobyl . . . conferred respectability on the idea of inherently safe reactors. If nuclear power was to survive, let alone contribute seriously to amelioration of

Table V. Nuclear Plants under Construction as of December 31, 1990

Country	Number of Units	Capacity (GWe)
Former Soviet Union	25[a]	21.3
Japan	10	9.0
France	6	8.3

NOTE: No other country had at least 4 GWe under construction.
[a]At least three of these were cancelled after the independence movement.
SOURCE: Data are from references 4 and 5.

the greenhouse effect, something new and different, something that would overcome the public's distaste for nuclear power was needed . . . the public, or at least the skeptical elites who influence public opinion, must be convinced by the transparency of a design of a device that the use of that device cannot harm the public.

The following sections briefly describe four classes of new designs.

Evolutionary Light-Water Reactors. The dominant type of nuclear reactor in the world is based on using regular water (light, to distinguish it from heavy water, used in some reactors) both to slow down (moderate) the neutrons, which cause and transmit the fission reactions, and to transfer heat and cool the reactor. The evolutionary reactors are large, 1300 MW (about the size of the largest U.S. reactors), and are based upon designs that have been built many times. Three of these evolutionary designs were discussed for possible use in the United States:

1. The one farthest along is the advanced boiling water reactor (ABWR) being developed as a joint venture by General Electric, Hitachi, Toshiba, and a group of Japanese utilities, under the leadership of the Tokyo Electric Power Company (TEPCO), which already has 13 operating reactors and three under construction. Two of these evolutionary BWRs were ordered, and one is already under construction.

2. Asea Brown Bovari Combustion Engineering, formerly Combustion Engineering, has developed an evolutionary version of their last plant, a pressurized light-water reactor (LWR). Two units of the new design, the System 80+, have been ordered by South Korea.

3. The advanced pressurized water reactor (APWR) is an evolutionary design being developed by a joint team from Westinghouse, Mitsubishi, and a group of Japanese utilities, led by Kansai Electric Company. Kansai has ten operating reactors and one under construction.

These three designs differ from current LWRs, but the differences are readily apparent only to designers, analysts, and, the vendors hope, to utility managers. All three plants are designed to be easier to build and operate, leading to shorter construction times, lower costs, and better operating performance. These plants also are designed to be safer, a characteristic that should be amenable to analysis. However, to the general public, and perhaps to investors and utility commissions, the differences from current designs are likely to be viewed as unimportant, which the term "evolutionary" may imply.

Smaller Reactors.   Beginning at least as early as the mid-1970s, some nuclear industry planners questioned the push for ever-larger nuclear plants. The argument for increasing the size was that doing so would bring "economies of scale", for example, that the amount of materials, design work, and construction workers would not go up linearly with plant size. Therefore, the cost per unit of capacity would decrease. For example, supporters of economies of scale believed that, if the plant size were doubled, the total cost would go up by less, perhaps much less, than a factor of 2.

Some doubt was expressed by critics, who raised three issues:

1. Was enough experience gained from midsized plants to warrant going to the larger plants or should the industry wait 5–10 years to gain experience from building and running 500–600 MWe plants before moving on to plants twice that size?

2. The larger plants took longer to build and were more expensive in total cost. Could utilities see the future well enough to predict when a large addition to generating capacity would be needed when that need would be at least 8, and perhaps longer, years in the future? Would a smaller unit, able to be built in a shorter time, be more likely to be needed when completed?

3. In other countries, particularly developing countries, was it more likely that smaller, and hence cheaper, plants would be needed rather than the very large, and much more costly, designs?

These questions did not receive serious attention until the past several years, when vendors and others in the industry concluded something significantly different might lead to new orders.

Although many may take credit for the examination of smaller designs, credit should be given to Juan Eibenshutz and John Taylor. Eibenshutz, then deputy director of the Mexican government utility (Subdirector, Commission Federal de Electricidad), in the early 1980s began to visit IAEA (International Atomic Energy Agency) member countries with the hopes of mounting an international effort to design and introduce a 200–300 MWe reactor. He argued that this size would be affordable, buildable, and operable in developing countries and would fit into the small grids of these countries. Although he received only lukewarm support and no financial backing, Eibenshutz did plant the seed for the idea that a smaller reactor might be attractive for countries not normally seen as potential markets.

Taylor directs the Nuclear Power Division of the Electric Power Research Institute (EPRI), the research arm of the U.S. electric utility industry. At about the same time Eibenshutz was arguing small reactors were marketable, EPRI was concluding that the evidence did not show support for economies of scale. Coupling that indication with the difficulty of predicting future capacity needs, Taylor convinced EPRI and the vendors to jointly sponsor development of smaller PWR and BWR designs, each about 600 MWe. The Department of Energy also is a cosponsor, along with utilities from six other countries.

The Westinghouse AP600 is the closest to completion. However, the following estimates are still of a design, not of a pilot or a full-size built plant. The AP600 design estimates, compared with a conventional 600 MWe plant, indicate the AP600 will have 60% fewer valves, 35% fewer large pumps, 75% less piping, 80% less control cabling, and 80% less ducting.

The General Electric small boiling water reactor (SBWR) also is estimated to have significant reductions from a conventional 600 MWe BWR, with 80% fewer fans, 73% fewer large pumps, and 16% fewer valves.

The major change in both of these designs is the introduction of a different approach to cooling the reactor in case of an accident. The normal method, including the approach taken for the evolutionary designs, is to rely on large pumps to force water into the reactor core to cool the fuel. (The Three Mile Island accident progressed from a difficulty to a disaster when these pumps were turned off and the water left the core.) In these new, small designs, cooling is accomplished by natural circulation. Basically, a large pool of water is placed above the reactor. Gravity is used to drive the water into the reactor, and the buoyancy of heated water is used to maintain circulation. Air flow, also natural, is used to provide cooling to carry the heat away. Much of the savings cited previously come from eliminating the systems to force the cooling water through the reactor.

If these designs are proven to work, they may be an example of necessity driving invention. A plausible (and attributed) explanation for the development is that designers were asked to reduce the capacity significantly but to keep costs proportional to size. Designers were asked to design around economies of scale, that is, to design a plant half the size of a big plant that also would cost only half as much. The approach decided upon was to eliminate many pumps, valves, ducts, and controls. Systems needed to operate the reactor could not be eliminated, leaving as the only candidates those systems used for safety shutdown. But if the safety systems were eliminated, how could the reactor be shut down? This led designers to the concept of using natural forces, the "passively safe" concept. The reactor size was then determined by how large the designers could make the reactor and still count on passive safety.

Many questions remain about these designs, including the following:

1. Can they be licensed in the United States without major changes? Elimination of active safety systems poses a major issue to the NRC. The NRC will review these designs to determine if they can withstand possible accident sequences. The NRC may conclude that some active safety systems will be required. If so both the savings and the concept of something new may be compromised.

   A possible delay may be introduced by the NRC's need to verify some of the performance claims by actual test, in its own facility. A November 14, 1991, Advisory Committee on Reactor Safeguards (ACRS) letter raised several concerns about what is necessary for NRC approval of passive designs. The NRC staff is considering whether the NRC must construct its own test facilities to model the AP600 plant (8).

2. Can the cost and performance goals be met? Unless pilot plants are built and operated, these questions will remain open.

3. Are the advantages sufficient to convince a utility to buy a plant? Obviously, this is the critical question. So far, the answer is no.

Significantly Different Water Reactors. There are several different water reactors. The two most discussed are the following:

1. The process inherent ultimately safe (PIUS) reactor is a 640 MWe Swedish design by Asea Brown Bovari Atom. Designed in response to the Swedish referendum to close down the Swedish nuclear program because of safety concerns, the PIUS' core is surrounded by a huge tank of water, with the water kept out of the core by a thermal barrier. The coolant water is physically in contact with the operating water at all times. An upset in the

system leads to flooding of the core with coolant. No serious interest in this reactor exists except for that of reactor analysts.

2. The Canadian deuterium uranium (CANDU) reactor is a Canadian design, which is used for all 20 of Canada's operating reactors plus the two under construction. The reactor uses heavy water (therefore, the deuterium), natural (i.e., unenriched) uranium, and on-line refueling. There are several other distinguishing features of these reactors. The Canadian designer, Atomic Energy of Canada, Ltd. (AECL), proposed an evolutionary design at 450 MWe. The original plan was to sell these in Canada, to the provinces currently without nuclear power, and also to market the design in the United States. Both plans are moving slowly, at best.

Other Concepts.   Two additional concepts of significant interest are as follows:

1. The high-temperature gas reactor (HTGR) has many strong supporters, at least in the United States. Also known as the MHTGR, where M indicates modular, this reactor has been discussed as being built at units of 135–175 MWe, although a recent suggestion is for units of over 200 MWe. This design features a graphite core, in which the fuel is located, and uses helium as the coolant. Among other features the reactor can operate at much higher temperatures than LWRs, leading to the possibility of providing process steam (540 °C or higher). The most notable feature, however, is that the safety is predicated upon "containment-in-a-pellet". The fuel is encased in small (0.8-mm diam.) pellets, with uranium oxide at the center and successive layers of pyrolytic carbon, silicon carbide, and pyrolytic carbon. The theory is that, even in the case of loss of all coolant, insufficient heat is generated by the amount of fuel in the pellet to melt or degrade the coatings. Hence, no radioactive material can escape.

   Many issues remain to be addressed regarding this reactor, including NRC licensing without a containment, ability to manufacture the pellets to the necessary quality, cost of the plant, and whether there are scenarios that could lead to radiation release. Economics may be the biggest problem. Nevertheless, this reactor has a growing band of ardent supporters, who see it offering a truly "inherently safe" design.

2. The liquid metal reactor (LMR) differs from all other types of reactors in both coolant and operation. The coolant is sodium and the neutrons used do not slow down before they cause fissioning. This feature is the reason this type of reactor is called a fast reactor. Using fast neutrons enables the reactor to breed.

Normal uranium contains mostly ^{238}U and a small amount of ^{235}U. The Canadian reactor uses natural uranium, which usually is about 0.7% ^{235}U. Most LWRs use enriched uranium, in which the percent of ^{235}U is about 3%. The ^{235}U fissions, while the ^{238}U does not. Hence, power generation comes from the ^{235}U. The Canadian reactor operates on so low a percentage of ^{235}U by using the improved properties of heavy water, which is about 300 times more efficient than light water for slowing down the neutrons. In the LMR the fast neutrons interact with the ^{238}U as well as the ^{235}U. Whereas slowed down neutrons interact well with ^{235}U, only higher energy neutrons interact with ^{238}U. (^{235}U fissions with thermal neutrons; the reaction ^{238}U plus neutron going to ^{239}Pu requires more than 1-MeV neutrons.) The ^{238}U is transmuted into plutonium, which then also fissions. Thus, the reactor breeds fuel and is known as a breeder reactor.

The LMR also offers improved safety characteristics, because it operates at low pressures and with a very efficient coolant. The principal advantage, however, is the ability to use the ^{238}U, thereby increasing the available fissionable resources by about 100 times. The major questions with respect to the LMR are economics and sensitivity to proliferation.

Table VI. Reactor Characteristics

Reactor	Safety	Economics	Market	Development	Licensing
ABWR	High	High	High	High	High
APWR	High	High	High	High−	High
80+	High	High	High	High−	High
AP600	(High)	(High)	Medium	Medium	Medium
SBWR	(High)	(High)	Medium	Medium	Medium
CANDU	High	High	Low	Medium	Low
PIUS	(High)	?	Low	Low	Low
HTGR	(High)	?	?	Medium−	Medium
LMR	High	?	Low	Medium	Medium+

NOTE: Parentheses indicate more uncertainty in the estimate.

Table VI gives a simplified summary of the relative characteristics of these reactors. The measures are my opinions based on a recent National Academy Report (9).

Future U.S. Nuclear Plants

Both widespread use of nuclear power and new designs exist. The United States uses nuclear power on a large scale, and plants are getting older. Why are new nuclear plants not being ordered? For plant orders there must be a need. Utility planners must see a need for new, large generating plants. Nuclear power must be seen as a good and wise choice by utility planners. Money must be raised for new plants. Therefore, nuclear power also must be seen as a good and wise choice by whoever finances these plants. For most U.S. utilities this group is the financial community and large stockholders. The plants must be approved, explicitly by the state regulatory commission and the NRC and implicitly by the public.

The final question is whether new nuclear plants will be ordered in the United States. To answer this question an understanding of why plants have not been ordered is required. Several factors are involved.

Demand for Electricity. In the early 1970s the Atomic Energy Commission estimated that over 1000 large nuclear plants would be in operation in the United States by the year 2000. The number is likely to be about 110, and many of these will be smaller plants. Many utility planners shared this estimate of 1000, which was predicated upon an electricity demand growing at about 7% per year. The first oil shock, in 1973, caused fundamental shifts in energy consciousness, energy use, and electricity growth. In 1974 the 10-year average annual growth rate for electricity use still was predicted to be 7.6%. In reality it was 2.9%. By 1978 some appreciation of the reduction in demand could be seen in plant cancellations and stretch-outs, and the estimated 10-year average was down to 5.2% annual growth; but, in reality, it was only 2.3%. Current estimates for the next 10–20 years range from 1.5 to 2.5%.

Many utilities over-built with plants that were started before the oil shock. This led to U.S. national reserve margins of as high as 27% by 1979. Of even more interest to utility planners was the growing use of prudency reviews by state regulatory commissions. Already by the late 1980s, some plants were becoming well known within the industry for the results of these reviews and for disallowances (money spent to build a reactor but that the rate commission refused to allow the utility to recover from ratepayers). Some examples are as follows:

Waterford, $260 million; San Onofre, $340 million; Limerick, $370 million; Callaway, $380 million; and Wolf Creek, $640 million. Others would reach over a billion dollars. By the end of the 1980s, nuclear plant capital disallowances approached $14 billion. Not lost on utility planners was that, over the same period, capital disallowances for non-nuclear plants totaled about $700 million.

The message is clear: Be very sure that a plant will be needed when it is completed or be prepared to have the stockholders suffer.

Construction. Unfortunately, construction time has not been well predicted or controlled. Part of the costs of a nuclear plant are the carrying charges while the plant is being constructed. A lengthy time adds to these charges and makes it more difficult for the planners to estimate correctly if the plant will be needed when ready. Table VII compares U.S. construction times with those of the other major nuclear countries. The two time periods could be labeled "Pre-Three Mile Island" and "Post-Three Mile Island". I have not included data from the former Soviet Union, because I know very little about the accuracy of such data.

Masked in these numbers is a large variation in the U.S. plant data. Although the average U.S. plant in the 1980s took about 12 years to finish, some took as many as 19 and a few only about 6. A utility planner would be hard-pressed to make a defendable estimate of when a new plant would come on-line.

Costs. The costs also were difficult to predict, but it was safe to estimate that they would be large. The usual method of comparing capital costs for electricity-generating plants is in dollars per kilowatt. Since most modern plants are slightly larger than 1000 MWe, the numbers in dollars per kilowatt are approximately total costs in millions. (Thus, at $1000/kw, a 1000 MWe plant would cost 1000 million,

Table VII. Average Construction Time

Country	1967–1978		1979–1990	
	Number of Units	Average Months	Number of Units	Average Months
Japan	20	51	20	56
France	10	69	48	72
Canada	10	85	12	94
Germany	17	58	15	99
United States	66	69	48	139
United Kingdom	10	90	10	153

NOTE: Measured from first pouring of concrete to connection to the grid.
SOURCE: Data are from reference 4.

or 1 billion, dollars.) Table VIII gives some costs that were reported in the trade press or by the utilities.

As can be seen the total costs have reached the 4- and 5-billion-dollar range. Outside of the Defense Department, these are sobering costs for a piece of technology. Some of this cost growth is related to general inflation in the U.S. economy, and some is due to the lengthening of construction time. However, these factors can be somewhat removed by estimating "overnight costs", a term used to describe the costs if there are no time charges. In constant 1988 dollars, to remove the inflation effect, the average overnight cost of a 1000+ MWe U.S. plant went from $1730/kw in 1981–1984 to $3100/kw in 1987–1988. Confounding the utility planner was the fact that the costs were not predictable: when the average was $1730, the range of lowest-to-highest-cost plant was from $1300 to $4200. When the average was $3100, the range was from $1400 to $4600.

Furthermore, for plants in operation during the entire decade of the 1980s, nonfuel operating and maintenance costs for nuclear plants rose 165%; for coal plants, these costs rose 38%.

Taylor, an ardent nuclear advocate, wrote of the effect of these costs (*10*):

> The rapid escalation of construction costs of those nuclear plants completed in the United States [since the oil shocks] has made them uneconomic compared to coal plants at present-day coal prices . . . In addition, the opportunity to counter the high capital costs with high operating capacity factors and low operation and maintenance costs was missed in the United States. Although some U.S. plants have

Table VIII. Announced Costs of U.S. Nuclear Plants

Plant	Year of Commercial Operation	Capital Costs ($/kw) (Current $)
Susquehanna 2	1984	1620
Catawba 2	1986	1630
San Onofre 2 and 3	1983, 1984	2050
Braidwood 1 and 2	1988	2280
Waterford	1985	2430
Millstone 3	1986	3300
Hope Creek	1986	4030
Fermi 2	1988	4220
Perry	1987	4260
River Bend	1986	4360
Vogtle 1	1987	5160
Beaver Valley 2	1987	5300
Nine Mile Point 2	1988	5830

operated as well as any in the world, others have experienced poor performance.

The performances were quite poor, in fact, and were probably caused by inferior management.

Performance. The role of a nuclear plant is to generate electricity. The more the plant can run, the lower the price of its electricity, because the capital cost and plant staff costs can be spread over more kilowatt-hours. The standard measure of performance is load factor, the percentage of time the plant is generating electricity at its full capacity. In the United States all nuclear plants are run as baseload plants, so that if they can run, they do run. France, with its large nuclear capacity, does use some of its plants in a load-following mode, so that they are not running at full capacity even when they can. This practice reduced the load-factor for France in the late 1980s. Table IX presents the lifetime load factors for major nuclear countries.

As can be seen the U.S. performance has been poor. Why? Is it unmerciful regulation? Is it the lack of a few huge utilities, or one government utility? Is it too many designs? In 1990 two nuclear engineering professors from Massachusetts Institute of Technology wrote (*12*):

> It is notable that LWR plants in the U.S. have been able to supply electricity to market for a lower proportion of their operating time than has been the case in many other countries. A 1986 study . . . found that the disparities result from differences in management and professionalism at individual plants, rather than in political or industrial structure. Some U.S. plants performed as well as any in the world, but others performed poorly enough to drag down the av-

Table IX. Lifetime Nuclear Power Plant Load Factors through 1987

Country	Load Factor (percent)
Canada	78.2
West Germany	73.6
Sweden	71.1
United Kingdom	69.1
France	68.1
Japan	68.0
United States	60.5

SOURCE: Data are from reference 11.

erage. It seems that vigorous anticipation of problems and attention to detail are key to good plant performance. Good management, therefore, is as important to the success of nuclear power as design is.

The performance of U.S. plants is improving: in the past 3 years, performance was 65, 63, and 68%. However, as Table X indicates, U.S. performance still lags behind that of many other countries (Table X gives the median, not the average).

Although there was considerable rearranging of rankings between 1984–1985 and 1987–1989, the United States remained in 13th place. As will often be noted by the nuclear industry, many of the world's best-running plants are in the United States. That is true. Over the period 1987–1989, of the top ten world nuclear plant load factors, eight were for U.S. plants (*13*). This fact merely highlights that, although some U.S. plants run very well, a much larger number do not.

Nuclear power advocates thus find it difficult to answer the three basic questions that an electric utility planner asks about a proposed new plant: (1) How much will it cost? (2) When will it be ready? and (3) How well will it run?

Public Opinion.   The final obstacle to nuclear power is public opposition. Polls can be quoted by both sides of the debate. A recent

Table X. Median Load Factors

1987–1989		Load Factor (%)	1984–1986
Rank	*Country*		*Rank*
1	Finland	90.8	1
2	Switzerland	84.0	2
3	Spain	83.0	14
4	Canada	82.6	7
5	Belgium	82.5	3
6	South Korea	78.6	11
7	West Germany	78.3	4
8	Sweden	77.0	8
9	Czechoslovakia	76.3	9
10	Bulgaria	75.7	6
11	Japan	73.4	12
12	United Kingdom	72.4	5
13	United States	69.4	13
14	USSR	68.8	NA
15	France	65.2	10

SOURCE: Data are from reference 13.

276 RADIATION AND PUBLIC PERCEPTION

cover story in *Time* focused on nuclear power: "Time to Choose". *Time* reported on two polls.

One asked "Which one of these energy sources should the U.S. rely on most for its increased energy needs in the next ten years?" The choices were nuclear, oil, coal, and other. These choices were not restricted to electricity use. The vote was 40% for nuclear, with the remaining split between oil and coal. Only 5% chose "other".

The second question was "Do you favor or oppose building more nuclear power plants in this country?" Thirty-two percent of the respondents were strongly opposed, and 20% were somewhat opposed.

The public seems to be of two minds. However, one must look beyond the questions. Other polls have shown that the public is unconcerned about energy.

For example (*14*):

> The number mentioning energy as one of the two most important problems [in the United States] dropped from 69% in 1979 at the height of the oil crisis to 1% or 0% [in 1991]. As of [fall, 1990], only 24% of Americans thought any new generating capacity would be needed in the next 10 years.

The American public is convinced that energy supplies are plentiful and that electricity will be available when needed. Hence, when they are asked questions about choices of types of energy, Americans are being asked to address topics they have already indicated are not of much interest to them. Consequently, little weight should be given to energy options, either for or against. On the other hand, questions relating to siting of facilities, whether they be power plants, industrial facilities, hazardous waste dumps, incinerators, or radioactive waste sites, do prompt reactions based on strong interest. The public does not want such facilities near them. Many acronyms have been formed to indicate these attitudes:

- LULU—locally unwanted land uses;
- NIMBY—not in my backyard;
- NIABY—not in anyone's backyard;
- NIMTO—not in my term of office;

and so on.

Nuclear power is afflicted with this problem. Public opposition is heightened by any accident, anywhere. The fact that Chernobyl was an accident in the former Soviet Union with a completely different type of reactor and a basically unregulated workforce is not significant to the public, who views it as an accident involving a nuclear reactor.

Operators asleep in control rooms, plants shut down because management refuses to take corrective actions, the Tennessee Valley Authority's (TVA) nuclear program halted for years, the brittle failure questions about Connecticut Yankee—all these occurrences keep in front of the public the idea that nuclear power abounds with problems. The costs associated with some recent plants and accompanying rate shocks when the plants come on-line exacerbate nuclear power's difficulties.

Nuclear power also is directly related to radiation—correctly so. However, radiation is poorly understood and viewed as mysterious and dangerous by most of the public. Many opposition groups use this fear as a key part of the argument against anything nuclear. Thus, in 1989 a booklet published by a group opposed to nuclear-waste sites (15) described radiation as follows:

> Prominent radiation experts, national and international scientific groups, and all U.S. government agencies agree essentially that no level of radiation is safe . . . A 1979 National Academy of Sciences (NAS) report said that virtually every type of cancer—blood, breast, lung, digestive system, and others—can be initiated by radiation exposure. NAS said that research also has linked heart disease, aplastic anemia, cataracts, shortened life span, and weakening of the immune response system to radiation exposure.

Perhaps the growing concern about greenhouse warming, and its relationship to CO_2 from burning fossil fuels, may lead to renewed interest in nuclear power? Last year a leading U.S. utility executive asked for a comparison of the waste from nuclear and coal (16):

> Let's compare two waste alternatives. A 1,000 MW coal-fired generating unit will produce annually 3.5 million cubic feet of ash; 35,000 tons of SO_2 at emission limits of the [amended] Clean Air Act; and 4.5 million tons of CO_2. A similar size nuclear unit will produce 70 cubic feet of high-level radioactive vitrified waste. The latter can be stored in a deep underground repository in a stable geologic formation. We can monitor it. But we don't know where the gaseous wastes from coal burning will go or what will be their long-term effects . . .

However, these arguments have not convinced utility executives to order nuclear plants nor have they been sufficiently convincing to the opponents (6): "Is there any realistic role for nuclear power in preventing climate disruption? Not likely, in Audubon's opinion."

Even the National Research Council is dubious. In a recent study on greenhouse warming, a committee examined many options for mitigating greenhouse effects. Regarding nuclear power, the report (17) states:

> Questions about the appropriateness of current technologies and public opposition to nuclear power, however, currently make this option difficult to implement. To the extent that concern about greenhouse warming replaces concern about nuclear energy and "inherently safe" nuclear plants are developed, this option increases in priority ranking.

Some additional points that may have significance in nuclear power's future are discussed in the following paragraphs.

Much of the world is poor, and most of the world's population is struggling for a better life. Last year, William Draper, the administrator of the U.N. Development Program, made these remarks (18):

> . . . the daily battles for survival waged by people of the developing world. It is here, in the teeming slums of Rio and Calcutta, in isolated villages in Mali and Niger, and in the devastated plains of Bangladesh that the future is being born.
>
> Every day, a quarter of a million people are added to the planet. Over 90 percent are born in the developing countries. Between now and the year 2000, world population is expected to grow by more than a billion people— the size of present-day China. Yet the world cannot adequately care for those who are here today. One person in every five lives in absolute poverty . . . One in three children is seriously malnourished.

The World Bank estimates world population will reach 8.5 billion by the year 2025. Eight countries are estimated to have more than 200 million people: USSR, the United States, China, India, Bangladesh, Nigeria, Pakistan, Indonesia, and Brazil (19). In 1985 over 1.1 billion people had an annual income of less than $370 (19).

As population grows there also is a migration into megacities. In 1950 there were 78 cities with a population of more than 1 million.

In 1990 there were 298. The U.N. predicts 639 such cities by the year 2025.

Chauncey Starr has shown a strong correlation between gross domestic product and electricity generation for both the developed countries and the less developed countries (LDCs) (*20*).

In the United States interest in nuclear power has been reviving. Three headlines from *The New York Times* over the past 2 years indicate this interest. These headlines are in chronological order.

1. "The Nuclear Industry Tries Again: Reactor Makers Promise Safer, Cheaper Designs as Fears Mount over the Greenhouse Effect" (*21*), obviously skeptical.

2. "Revive the Atom" (*22*), supporting passively safe plants.

3. "Reviving Nuclear Power from Its Coma" (*23*), supporting evolutionary plants.

Summary

Also, electricity use continues to grow: 36 utilities set records for summer peak demand in 1991.

For new nuclear power plants to be ordered in this country, the following must occur:

- The demand for electricity must be greater than can be met by conservation, load management, and renewable energy sources.

- Current nuclear plants must operate without accidents that lead to major releases of radiation or loss of a reactor.

Nuclear proponents often seem to forget that electric utilities are in business to provide reliable, low-cost electricity. The source of that electricity is of little concern to the consumer and of increasingly little concern to the utility. In the future utilities will move toward being distribution and transmission companies and will purchase electricity from the most cost-efficient source, including conservation (so-called "negawatts") (*24*). Therefore, the following also must occur:

- Nuclear power must be seen as economically competitive. This turnabout can happen if fossil fuels are priced out of the market by a carbon tax, for example, or if new nuclear designs are perceived to have significantly lower costs.

- The utility industry must be convinced that new plants, unlike the current generation of plants, will be built within a known, and shorter, time; will come on-line at a known, and lower, cost; and will run better.

These differences do not require major new designs. Current plants that are well managed and operate efficiently and the successful introduction of evolutionary designs could satisfy all the requirements, if fossil fuels were constrained.

Many people in the industry believe that new plants must be seen by the public to be much safer than current plants and that a waste disposal site must be chosen and, preferably, be under construction.

In conclusion,

- There are new designs for nuclear power plants. Although they exist mostly on paper, they *are* new.

- Nuclear power is very important in many countries and in many states.

- Good operation of nuclear plants and constraints on fossil fuel use are needed in the United States for nuclear power to recover.

My concern remains with management. Some utilities do very well, but many do not.

The only publicly expressed interest by a U.S. utility executive in nuclear power in the past few years was by the chairman of the TVA. The TVA sets its own rates, is publicly financed, and has had a miserable record in building and operating a large nuclear program. Recently, a more cautious tone was evident in an article by TVA's chief financial officer: ". . . the design and construction of nuclear power plants continue to present planners with troublesome risks. Such risks make a new nuclear power project financially undesirable at this time" (25)[1].

I am not optimistic.

References

1. Barnert, H.; Krett, V.; Kupitz, J. *IAEA Bull.* 1991, 33(1), 21–24.
2. Lu, Yingshon *Nuclear District Heating Reactor Development in China*; Institute for Techno-Economics and Energy System Analysis: Beijing,

[1]The TVA notified the NRC that the former does plan to resume construction at Bellefonte 1 in Alabama, although no date was specified. Bellefonte was halted in 1988 when 85% complete. Completion of the 1260 MWe plant is estimated to cost $1.5–2 billion (26).

China, undated. Wang, D.; Changren, M.; Jiaguin, L. *A 5-MW Nuclear Heating Reactor*; ANS Transaction Proc. 7th Pacific Basin Nuclear Conf.; March 4–8, 1990.

3. *Annual Energy Review 1990*; DOE/EIA-0384(90); May 1991; p 285, Table 127.
4. IAEA Ref. Data Series No. 2, 1991.
5. BP Statistical Review of World Energy, 1991.
6. Beyea, J. Testimony before House Subcommittee on Energy and Power; 15 March 1989.
7. Forsberg, C. W.; Weinberg, A. M. *Ann. Rev. Energy* **1990**, *15*, 133–152.
8. Nov. 14, 1991, letter from David Ward, ACRS Chairman, to Ivan Selin, NRC Chairman.
9. *Nuclear Power: Technical and Institutional Options for the Future*; National Academy Press: Washington, DC, 1992.
10. Taylor, J. H. *Science (Washington, D.C.)* **1989**, *244*, 318.
11. IAEA PRIS Report NBLG020G 89-02-02.
12. Golay, M. W.; Todreas, N. E. *Sci. Am.* **1990**, *262*, 84–85.
13. *Nucl. News* **1991**, *34*(3), 44.
14. Bisconti, A. S. *Public Opinion in the United States and Canada: Five Reasons for Cautious Optimism*; Workshop on Nuclear Energy Public Information Policies and Programs in OECD Countries; March 1990; pp 1–2.
15. *The Workbench*; Southwest Research and Information Center, 1989; Vol. 24(2), p 48.
16. Lee, W. S. *Energy for Our Globe's People*; 5th Annual Emerging Issues Forum; North Carolina State University: Raleigh, NC, February 9, 1990.
17. *Policy Implications of Greenhouse Warming*; National Academy Press: Washington, DC, 1991; p 231.
18. Draper, W. H., III National Press Club: Washington, DC, May 22, 1991.
19. *World Development Report 1990*, Oxford University Press: Oxford, United Kingdom, 1990; pp 29 and 228–229.
20. Starr, C.; Searl, M. F. *Energy Systems and Policy* **1990**, *14*, 53–83.
21. *The New York Times*, Nov. 26, 1989.
22. *The New York Times*, editorial, Dec. 8, 1989.
23. *The New York Times*, editorial, Nov. 1990.
24. Willrich, M. *Public Utilities Fortnightly*, Oct. 1, 1991.
25. *TVA Re-Examines the Nuclear Option*; Forum for Applied Research and Public Policy, Winter 1991; pp 87–90.
26. *Nuclear Energy Info*; USCEA: Washington, DC, April 1993; p 8.

RECEIVED for review January 14, 1993. ACCEPTED revised manuscript April 29, 1993.

Science, Society, and U.S. Nuclear Waste

Ginger P. King

Education and Information Division, Office of Civilian Radioactive Waste Management, U.S. Department of Energy, 1000 Independence Avenue, SW, Washington, DC 20585

This chapter discusses nuclear waste generation, its characteristics and location; the components of the waste-management system; the characteristics of a permanent geologic repository for spent fuel and high-level radioactive waste; and the scientific studies and activities required to develop a waste-management system for safe disposal of spent fuel and high-level radioactive waste, emphasizing particularly the importance of geotechnical and geochemical issues. This chapter presents a general overview of the U.S. high-level radioactive waste-management program as mandated by the Nuclear Waste Policy Act of 1982, as amended, and further discusses its associated task of integrating scientific conclusions with societal concerns.

T HE UNITED STATES AND ALL OTHER MAJOR nuclear electricity-generating countries have policies and plans for the permanent isolation, or disposal, of spent nuclear fuel and high-level radioactive waste in deep geologic repositories. Scientific consensus determined that this method is the safest, most desirable means for protecting the public health and the environment. The development of such a waste-disposal system is not a simple task. Many complex scientific, technological, and societal issues must be addressed to establish the system. One of the biggest challenges in the development of a waste-management system is to effectively bridge the gap that exists in the nuclear field between scientific conclusions and societal acceptance of these conclusions. Effective, two-way communication with interested parties and affected governments is necessary to build the public's trust and confidence in this highly scientific program. Only when the public's trust is gained can the waste disposal system, mandated by the Nuclear Waste Policy Act of 1982 (1), as amended, be successfully implemented.

Commercial nuclear power and nuclear national defense activities produce radioactive spent fuel and high-level waste. This radioactive spent fuel and high-level waste is temporarily stored in 35 states (including Idaho, which currently stores spent fuel from the shutdown commercial reactor at Ft. St. Vrain near Platteville, Colorado) at the locations where they are generated, but so far no facility exists for the permanent disposal, or permanent isolation, of these materials. The lack of a permanent facility is not unique to the United States. Of the approximately 30 other countries with nuclear electricity-generating capacity and nuclear-defense activities, none has permanent disposal facilities. However, the United States and most of the other major nuclear producing countries have the same plans: to permanently dispose of spent fuel and high-level radioactive waste in deep geologic repositories.

Nuclear Waste: What Is It and Where Is It?

Nuclear waste comes from five major sources:

1. the steps involved in using nuclear energy to produce electricity, that is, the nuclear fuel cycle;

2. national defense activities;

3. hospitals, universities, and research laboratory activities;

4. industry; and

5. mining and milling of uranium ore.

And, from these sources, there are four basic types of nuclear waste:

1. spent fuel and high-level, or "long-lived," radioactive waste;

2. low-level radioactive waste;

3. transuranic, or what some countries may call intermediate-level, waste; and

4. mill tailings.

These classifications depend on the waste's source, level of radioactivity, and potential hazard.

While the volume of radioactive waste does not indicate its level of radioactivity, both the volume of waste and how much radioactivity it contains are important. For example, spent fuel is far less than 1% of the total volume of radioactive waste, but it contains about 96%. On the other hand, low-level radioactive waste adds up to nearly 86% of the total volume of radioactive waste but contains less than 0.1% of the total radioactivity of all radioactive waste (Figure 1).

Spent Fuel and High-Level Waste. High-level waste is the most radioactive of all nuclear wastes. It includes spent nuclear fuel (also referred to as spent fuel) from the nuclear generation of electricity, high-level "residue" from the reprocessing of spent fuel, and some wastes from the nation's nuclear defense activities. All of these are now in temporary "storage" awaiting permanent disposal.

Unlike many chemical substances spent fuel and high-level radioactive waste lose their radioactivity (or decay) over time. Some wastes decay quickly, while others may take a long time.

Spent fuel comes from nuclear electricity generation. The fuel for nuclear power plants is uranium oxide formed into ceramic pellets. Each pellet is about 3/8 in. (0.95 cm) in diameter and 1/2 in. (1.3 cm) long, about the size of the tip of a child's little finger. The pellets are stacked and sealed in fuel rods, hollow metal tubes about twice the thickness of a pencil and about 12 ft (3.6 m) long.

Groups of fuel rods are spaced and bolted together to form a fuel assembly. A fuel assembly contains about 200 fuel rods. These fuel assemblies are loaded into a reactor. The number of fuel assemblies varies and depends on the design of the reactor.

In the reactor the uranium atoms in nuclear fuel produce the energy needed for a nuclear power plant by fissioning, or splitting, into smaller atoms. In the process they release a great deal of heat, or electrical power. Fission is the process in which a uranium atom absorbs a neutron and then splits into two smaller atoms, releasing a relatively large amount of energy and one or two neutrons. These neutrons, in turn, can cause other uranium atoms to fission, releasing more energy and still more neutrons. Instantaneously, a nuclear reaction is achieved in which only one neutron from each uranium atom that fissions causes another uranium atom to fission. Such a nuclear reaction is a "nuclear chain reaction". The nuclear chain reaction produces the energy that is converted to electricity at a nuclear power plant.

Over time, as the reactor operates, the fuel becomes less efficient in its fissioning process. After about 3 to 4.5 years in the reactor, the

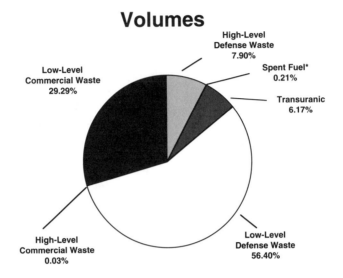

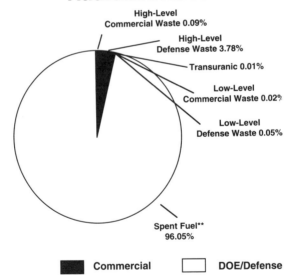

* Commercial LWR spent fuel permanently
  discharged. Includes spacing between fuel
  assembly rods.

**Commercial LWR spent fuel
  permanently discharged.

Figure 1. Radioactive wastes: volumes and radioactivities (1992). U.S. Department of Energy. (Reproduced from reference 5. This report provides historical data as of December 31, 1992. The figure was created in March 1994.)

fuel assemblies are removed from the reactor and defined as "spent fuel". As a result of the fissioning process, the fuel rods and their contents are very radioactive. Some of these fission products decay, or lose their radioactivity, within minutes or hours; others may take decades or centuries.

Based on 1992 data, approximately 28,312 metric tons of spent fuel will have been generated as a result of commercial nuclear power through 1993. Table I shows the amounts of spent fuel generation projected by state for 1993 and the amount projected by the year 2003.

After being removed from the reactor, the spent fuel is placed in specially treated water in deep, steel-lined concrete pools inside a building at the power plants. The water helps to thermally cool the fuel assemblies while the radioactivity decays, and the water, steel, and concrete in the pool shield employees and visitors from the radiation. Some utilities are now removing spent-fuel assemblies from the pools and placing them in dry storage on-site to provide additional storage capacity.

During the first 3 months of the spent fuel's storage, it loses 50% of its radioactivity. In 1 year it loses about 80%. In 20 years radioactivity is reduced by 90%. The remaining 10% forms the long-lived radioactive elements and is highly radioactive.

In the late 1960s and early 1970s, a spent fuel reprocessing industry began in many countries, including the United States. Commercial reprocessing plants were built at West Valley, New York, and Barnwell, South Carolina. However, in 1976 President Carter placed a moratorium on the reprocessing of commercial spent fuel because of the concern that the plutonium that is separated from the spent fuel by reprocessing would be accessible for nuclear weapons and thereby facilitate the proliferation of nuclear weapons. As a result of a brief reprocessing business at West Valley between 1966 and 1972 (no reprocessing occurred at Barnwell), there is a small amount of commercial high-level radioactive waste.

Although President Reagan removed the moratorium on reprocessing in 1981, today, since the uranium supply in the United States is plentiful, it is more cost effective to mine uranium than it is to reprocess the waste. In other words it takes less energy to fabricate and use fresh fuel than to separate the isotopes from the spent fuel for reuse. For this reason commercial spent fuel is not currently being reprocessed in the United States.

In many countries, such as France, Finland, Sweden, and Switzerland, utility companies are responsible for both temporary storage and permanent disposal of the spent fuel they produce. In the United States, while producers of spent fuel and high-level radioactive waste

Table I. Commercial Spent Fuel Storage, 1993 and 2003 (metric tons of uranium)

State	1993	2003
Alabama	1,334	2,130
Arizona	430	1,125
Arkansas	554	884
California	1,253	2,181
Colorado	15	15
Connecticut	1,189	1,752
Florida	1,320	2,048
Georgia	915	1,713
Idaho*	51	51
Illinois	4,154	6,701
Iowa	231	360
Kansas	194	420
Louisiana	318	790
Maine	426	580
Maryland	578	882
Massachusetts	431	489
Michigan	1,149	1,951
Minnesota	610	930
Mississippi	299	590
Missouri	242	470
Nebraska	350	610
New Hampshire	63	295
New Jersey	1,080	1,855
New York	1,792	2,588
North Carolina	1,460	2,311
Ohio	395	810
Oregon	358	359
Pennsylvania	2,284	4,250
South Carolina	1,684	2,923
Tennessee	409	1,066
Texas	320	1,155
Vermont	365	502
Virginia	1,088	1,804
Washington	191	363
Wisconsin	779	1,128
Total	28,312	48,045

SOURCE: U.S. Department of Energy. *Nuclear Fuel Data*; Form RW-859, 1992.
NOTES: The sums of the entries may not equal the totals due to rounding errors.
Compiled May 1994.
*No commercial reactor has operated in Idaho, but the Idaho National Engineering Laboratory currently stores spent fuel from the shutdown commercial reactor at Ft. St. Vrain, near Platteville, Colorado.

are responsible for their temporary storage, permanent disposal is the responsibility of the federal government. This responsibility was established by the Atomic Energy Act of 1954 (2, 3).

Because U.S. utilities are responsible for the temporary storage of the waste they produce, spent fuel and high-level radioactive waste from nuclear power plants are currently stored at reactor sites awaiting final, permanent disposal. These temporary storage sites are tightly controlled by the U.S. Nuclear Regulatory Commission (NRC), a federal agency.

As producer the U.S. government is responsible for both the temporary storage and permanent disposal of high-level radioactive waste from U.S. defense activities. This waste, measuring in volume about the equivalent of 9000 metric tons, is currently stored at three U.S. Department of Energy (DOE) sites: the Savannah River Plant in Aiken, South Carolina; the Idaho National Engineering Laboratory in Idaho Falls, Idaho; and the Hanford Reservation in Richland, Washington.

In 1982 the U.S. Congress mandated a national policy for developing the U.S. nuclear-waste-management system for the disposal of spent fuel and high-level radioactive waste. This national policy was established by the Nuclear Waste Policy Act (NWPA) of 1982 and was subsequently amended in 1987 to further focus the national program. The act created the Office of Civilian Radioactive Waste Management within the DOE to implement the policy and to develop, manage, and operate a safe waste-management system.

Passage of the NWPA was a major milestone in the nation's management of spent fuel and high-level radioactive waste. Since 1957, when the first U.S. commercial-generating nuclear power plant, a converted U.S. Naval reactor, began operating in Shippingport, Pennsylvania, studies for isolating spent fuel and high-level radioactive waste have been in progress. The National Academy of Sciences (NAS) in 1957 published results of a study which determined that the safest long-term means of disposing of spent fuel and high-level radioactive waste for the protection of the public health and the environment is in stable, deep geologic formations.

Despite this scientific finding U.S. policy and funding for pursuing such waste management and disposal solutions fluctuated: some administrations and congressional appropriations provided funding, while some did not.

Finally, on December 20, 1982, after years of Congressional debate and in the waning hours of the 91st Congress, the NWPA was passed. This law established a national policy for safely storing, transporting, and disposing of spent fuel and high-level radioactive waste. Recognizing that permanent disposal of spent fuel and high-level waste is both a scientific and societal challenge, the NWPA, as amended, requires a variety of mechanisms to ensure that affected governments and interested parties have extensive rights to oversee and participate in the program.

A key feature of the law is the establishment of the Nuclear Waste Fund, into which the producers of spent fuel and high-level radioactive waste must pay for the cost of developing the waste-management system. Commercial utilities pay 1 mill (one-tenth of a cent) per kilowatt hour into the fund for nuclear electricity generated and sold. Typically, these costs are passed on to utility customers. The federal government is required to pay into the Nuclear Waste Fund, because it produces U.S. defense high-level radioactive waste.

In the United States electricity is produced by 109 commercial nuclear power plants located in 32 states. Figure 2 shows the locations of nuclear power plants in the United States.

Low-Level Radioactive Waste.   Low-level radioactive waste usually contains only a small amount of radioactivity within a relatively large volume of material. Most low-level radioactive waste does not require extensive shielding from people and the environment, but some protective shielding may be needed for handling certain low-level waste.

Low-level radioactive wastes are generated by hospitals, laboratories, industrial plants, nuclear power plants, and government and

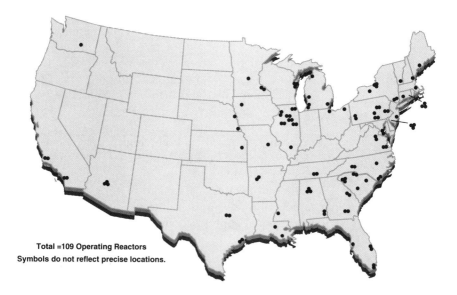

Total =109 Operating Reactors
Symbols do not reflect precise locations.

Figure 2.   Location of U.S. nuclear power reactors. U.S. Department of Energy and U.S. Nuclear Regulatory Commission. (This figure was created in May 1994.)

defense laboratories and reactors. These wastes include rags, papers, filters, resins, and discarded protective clothing. They are placed in containers, such as barrels or drums, which are then placed in shallow burial at special landfills licensed by the federal government. Two commercial low-level radioactive waste disposal sites are now in use in Barnwell, South Carolina, and Hanford, Washington. A third site in Beatty, Nevada, closed on December 31, 1992. Low-level radioactive wastes from national defense activities are disposed of at U.S. DOE sites.

In some countries, such as Finland, Sweden, and Switzerland, deep geologic disposal is used for low-level radioactive wastes. In the United States the policy for disposal of low-level radioactive waste was established by the Low-Level Radioactive Waste Policy Act of 1980 (4), as amended. This policy calls for each state, by January 1, 1996, to be responsible for providing for the safe disposal of the low-level radioactive waste produced within the state and for the federal government to take responsibility for any low-level radioactive waste resulting from federal activities. A state may form a "compact" with other states to provide a regional facility for all members of the compact to use. Congress must ratify, or approve, such compacts. Once approved a compact can refuse to accept low-level radioactive waste from nonmembers. If it prefers a state may operate its own disposal facility with the right to exclude waste from other states.

As of 1992 a number of compacts were ratified by Congress and some states established their own disposal policies for disposal of low-level radioactive wastes. Figure 3 shows existing compacts as of April 1993 and those states that chose not to join a compact.

Transuranic Wastes.   Like low-level radioactive waste, transuranic wastes are mostly clothing, rags, equipment, containers, and tools possibly "contaminated by" or containing radioactivity. Unlike low-level radioactive waste, transuranic wastes may contain elements with very long half-lives. This means that they lose their radioactivity slowly and remain radioactive for thousands of years.

Most transuranic wastes result from reprocessing nuclear fuel and making plutonium weapons. As discussed earlier commercial reprocessing of spent fuel is not currently practiced in the United States, although there is no prohibition against such activities. Reprocessing does occur in the national defense activities of the United States for the principal purpose of separating plutonium from spent fuel and then using the plutonium for the production of nuclear weapons as part of U.S. defense activities. The resulting wastes are now being stored at U.S. DOE facilities.

Figure 3. Regional compact status. U.S. Department of Energy. This figure was created in April 1993.

Plans call for transuranic wastes to be disposed of in a repository deep underground. The U.S. government plans to test geologic disposal of transuranic wastes at the Waste Isolation Pilot Plant (WIPP) facility in Carlsbad, New Mexico. This geologic disposal facility is carved out of salt more than 2000 ft (600 m) below ground.

Mill Tailings. The fuel used at nuclear power plants comes from uranium ore, which has been found in, and mined from, the ground. The uranium ore is mined and then "milled", that is, crushed and treated to separate and remove the uranium. The residue consisting of rocks and soil is called "mill tailings". These tailings contain a small amount of radium that decays to radon, a radioactive gas now recognized throughout the world. Radon can be harmful to the health of humans who are exposed to it in concentrated amounts. Mill tailings are covered with enough soil to prevent wind and water erosion as well as to prevent the release of radon.

The NWPA of 1982, as Amended

The NWPA laid out a process of and a schedule for identifying a site for a geologic repository for the permanent disposal, or isolation, of spent fuel and high-level radioactive waste. The law specified areas of scientific study to determine technical suitability of a candidate site and called for a comprehensive program known as "site characterization". Site characterization is a detailed process of scientific investigation, data collection, analysis, and evaluation. As defined by the law it is ". . . activities, whether in the laboratory or in the field, undertaken to establish the geologic conditions and the ranges of the parameters of a candidate site relevant to the location of a repository . . ." (5).

At the time of the passage of the 1982 law, the DOE was already conducting site studies on two federal locations: Hanford, Washington (6), and Yucca Mountain, Nevada (7, 8). In February 1983 the DOE carried out a requirement of the NWPA by formally identifying nine potentially acceptable sites for a permanent repository: one in Louisiana, two in Mississippi, two in Texas, two in Utah, and the two already under study (9–11). Amending the NWPA in 1987, Congress directed the DOE to conduct site characterization studies only at the Yucca Mountain site in Nevada as the candidate for a repository and to terminate site-specific studies at all other sites.

The purposes of site characterization are to obtain the information needed to determine as soon as possible whether Yucca Mountain is suitable for development of a repository, to acquire data necessary to develop more advanced designs for the potential repository and waste package, to conduct the quantitative evaluations or performance assessments needed to evaluate site suitability, and to demonstrate that a repository at the Yucca Mountain site, if it is found to be suitable, will comply with NRC requirements for licensing.

If the Yucca Mountain site is found suitable, the DOE must then demonstrate to the NRC that the site meets NRC regulations intended to protect the health and safety of the public both during operation of a repository and after the repository is no longer receiving spent fuel or high-level waste for disposal. In order to demonstrate to the NRC that the repository system at the candidate site will perform as required, designs must be developed for the repository, as well as for the waste package itself. The waste package consists of the waste (spent fuel or high-level radioactive waste) and the container in which it is packaged for disposal.

The Waste Management System

The U.S. waste-management system includes three elements: a mined geologic repository for permanent disposal of spent fuel and high-level radioactive waste, a temporary central federal storage facility (a monitored retrievable storage (MRS) facility, Figure 4) for temporary above-ground storage of commercial spent fuel, and a transportation system connecting all elements of the system.

Monitored Retrievable Storage. The earliest that a permanent, underground geologic repository will be available in the United States to begin receiving spent fuel and high-level radioactive waste for disposal is the year 2010. In the meantime commercial nuclear power plants need to keep pace with their ever-growing storage needs.

An MRS facility, authorized by the NWPA, as amended, can provide the storage space needed for commercial spent fuel until a repository is available and ready to begin receiving spent fuel and high-level radioactive waste for disposal.

Currently, several possible storage concepts are under consideration for an MRS facility. These include concrete and metal containers, multiple-element sealed metal canisters in concrete modules, modular vaults, and metal dual-purpose containers for transportation and storage. The DOE is also developing a multipurpose canister system, which could provide containers designed for transportation, storage, and disposal.

Storage of spent fuel in concrete containers on an open concrete pad is currently in use in the United States at Virginia Power's Surry Plant near Richmond, Virginia. This storage concept, licensed by the NRC, is also being used in other countries, such as Canada. These concrete storage containers are made of heavily reinforced concrete, and their walls are designed to provide radiation shielding. The containers have inner liners of steel and are equipped with racks for holding the spent fuel in place. They are stored vertically on a concrete floor in a storage yard. As all of the containers are independent storage units, each is modular. The capacity of these containers ranges from 24 to 32 spent fuel assemblies; their overall dimensions are 18 to 22 ft (5.4 to 6.6 m) in height and 11 to 12 ft (3.3 to 3.6 m) in diameter, with a loaded weight from 180 to 200 tons.

Another licensed dry-storage technology currently being used in the United States and also licensed by the NRC is a design in which spent fuel is placed in a sealed metal canister and placed horizontally in a concrete module. This dry storage technology is currently being used in the United States in South Carolina at Carolina Power and Light's H. B. Robinson Plant and at Duke Power's Oconee Plant. The canisters are made of stainless steel and have a storage capacity of 7 to 24 spent fuel assemblies.

The law provides for only limited amounts of spent fuel to be stored at an MRS facility and guarantees that the facility is not permanent. The law calls for a dual approach to siting the MRS. These two approaches are (1) siting by the DOE through a survey and evaluation

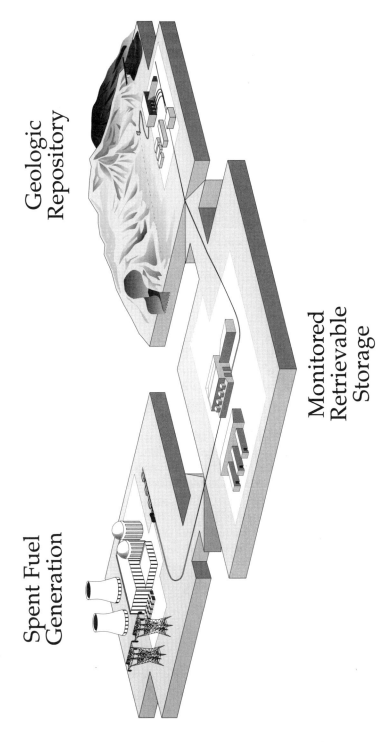

Figure 4. Waste management system with monitored retrievable storage facility. U.S. Department of Energy. This figure was created in August 1994.

process and (2) siting through the efforts of the U.S. Nuclear Waste Negotiator.

The negotiator, appointed by the president and confirmed by the U.S. Senate, is an independent federal position separate from the DOE. The negotiator's role is to seek a state or Indian tribe with a technically qualified site and to negotiate a proposed agreement on reasonable terms. Any proposed agreement is submitted to Congress for approval before becoming effective.

Geologic Repository. A geologic repository will resemble a large mining complex and combine two types of industrial facilities: a waste handling facility at the surface and an underground disposal facility built about 1000 ft (300 m) below the surface for the permanent disposal of waste in special containers.

Surface facilities will include waste-handling buildings, office buildings, fire and medical stations, water- and sewage-treatment plants, warehouses, repair and maintenance shops, a security office, and a visitor center. Shafts and ramps will connect the surface and underground areas.

Underground facilities will include main tunnels, called drifts, leading to the areas where the waste containers will be placed. The disposal area will consist of smaller tunnels with boreholes in the floor, and perhaps also in the walls, for the canisters of waste. Figure 5 is a conceptual drawing of what the repository might look like if located at Yucca Mountain.

The purpose of a geologic repository for spent fuel and high-level radioactive waste is to protect present and future generations and the environment from potential hazards of the radioactivity contained in spent fuel and high-level waste. United States Environmental Protection Agency (EPA) standards and NRC regulations require that the repository be designed to provide reasonable assurance that, during the 10,000 years of disposal, cumulative releases of radioactive isotopes to the environment will be kept within specific limits. In addition the waste packages must provide substantially complete containment of the waste for at least 300 to 1000 years. Thereafter, the waste package must be able to limit the rate of release of radionuclides to levels below the regulatory requirements for waste isolation.

The specific characteristic of the geologic repository that will allow the waste to be isolated for the required period of time is its multiple barrier design. The multiple barrier design, or system, includes both engineered (man-made) barriers and natural (geologic) barriers. It consists of (1) the waste package; (2) the repository itself; and (3) the "host rock", or geologic environment, in which the repository is built. The

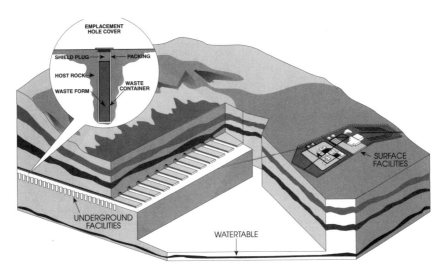

Figure 5. Conceptual drawing of geologic repository. U.S. Department of Energy. This figure was created in August 1994.

use of man-made and natural barriers is sometimes referred to as "defense in depth".

The waste package will be the principal engineered barrier in the repository's multiple barrier system. This package will consist of the waste form and a disposal container. Both spent fuel and defense high-level radioactive waste will be disposed of as solids, either in ceramic or glass form. No liquids will be disposed of in the repository. The disposal container will separate the waste from the host rock and be constructed of corrosion-resistant material.

The repository portion of the multiple-barrier system will consist of engineered barriers that are not part of the waste package. Material used to backfill (or refill) underground disposal rooms, passageways, ramps, and shafts will limit or control movement of underground water.

Natural, or geologic, barriers will have an important role in ensuring that waste will be isolated from the accessible environment when a repository is built. These barriers include the rock and the water, their chemistry, and how they might interact with the waste. For example, a crucial factor is groundwater movement: the time required for groundwater to flow from the repository to the accessible environment. Evidence of long-term geologic stability is desirable. Low and slow groundwater flow rates and long pathways from a repository to the accessible environment are desirable. In addition the ability of the rock to conduct heat away from the waste package is particularly important. Radioactive waste is thermally hot and would heat the waste package, as well as the surrounding rock. Because heated rock tends to drive away water, this would help to reduce any migration of nuclear waste into the environment. Rock properties that prevent or slow movement of harmful substances are also desirable. The site selected should also have a low possibility of human intrusion.

As mentioned earlier the Yucca Mountain site in Nevada has been designated by law for detailed study (site characterization) to determine as soon as possible whether or not it is suitable for a geologic repository for spent fuel and high-level radioactive waste.

If Yucca Mountain is determined suitable and approved for development as a repository, the repository complex will use about 5700 acres (2307 hectares) that will include a controlled area 3 miles (4.83 km) wide surrounding the outer perimeter. The surface facilities will probably cover from 150 to 400 acres (60.7 to 161.9 hectares). Gently sloping ramps connecting the underground and surface facilities will allow shielded transport vehicles to carry waste packages to the underground disposal area.

The underground facilities will cover an underground area of about 1400 acres (566.6 hectares) and be about 1000 ft (300 m) beneath the surface.

Transportation of Spent Fuel and High-Level Radioactive Waste. Nuclear materials, including spent fuel and high-level radioactive waste, have been safely transported for the past 40 or more years. Scientists and engineers worked together to design, test, and build the containers that the spent fuel and high-level radioactive waste are moved in to ensure the safety of the transporters, the public, and the environment.

The containers, or casks, used to transport spent fuel rods are designed and constructed to contain radioactivity under normal travel conditions and under severe, although unlikely, accident conditions by rail and highway. Safety tests conducted to certify transportation cask designs include a 30-ft (9-m) drop onto an unyielding surface; a drop onto a 6-in. (15.2-cm)-wide and 8-in. (20.3-cm)-tall steel rod; exposure to a 801.6 °C fire for 30 minutes; and submersion in water. In each case damage to the casks in computer modeling, scale testing, and full-sized tests proved to be superficial, and the simulated radioactive contents remained contained, unbreached, and isolated from the environment.

Site Characterization

As discussed earlier the purpose of site characterization at Yucca Mountain is to obtain the information necessary to determine whether it is a suitable site for a repository and, if so, to obtain NRC authorization needed to construct a repository. The information to be collected, analyzed, evaluated, and reviewed by scientists and the public will serve to establish (1) whether a repository can be constructed and operated at that site without adversely affecting the health and safety of the public during repository operations and (2) whether the waste emplaced in the repository will remain isolated from the general environment for thousands of years.

To determine whether the site is suitable, data are needed on the geologic, geoengineering, hydrologic, geochemical, climatological, and meteorological conditions at the site. These data will be obtained by investigations both on and below the surface. These investigations (briefly listed in Table II), their validation, and their review are expected to take a decade. As the repository program is a first-time effort, one of the related benefits of preparing for and conducting site characterization investigations is the expectation of advances in science and technology. One example of this possibility for scientific and technological advancement resulting from program studies is the development of the Exploratory Studies Facility (ESF). The ESF basically will be an underground laboratory to provide access to the potential host rock for a repository and evaluate conditions in the rock and the surrounding units. It uses a new, more efficient design than was used in the past for similar structures and was developed because of the site characterization program (Figure 6).

Surface-based investigations include tests performed both at the surface and in deep and shallow boreholes and trenches. Underground investigations require construction of shafts and ramps from the surface to below ground, where the repository would be located, so that people and equipment can fully explore and examine the geohydrology.

Detailed plans for conducting these investigations are included in a 6000-page Site Characterization Plan (SCP) for the Yucca Mountain Site. The SCP presents general information on the activities to be conducted at Yucca Mountain, the sequence of the activities, the priorities assigned to the activities, and the general schedules for the site-characterization program. Detailed descriptions of specific studies and activities are defined in study plans. Although the SCP does not include activities that will be performed to collect data on environmental and socioeconomic conditions, environmental and socioeconomic studies are also important and are required. These studies and activities are described in other documents.

Before discussing the geotechnical aspects that are involved in determining whether Yucca Mountain is suitable for a repository, it is important to understand a little about the history of the site-screening

Table II. Investigations To Be Conducted in the Site Program

Characterization Program	Investigation
Geohydrology	Regional hydrologic system
	Unsaturated-zone hydrologic system
	Saturated-zone hydrologic system
Geochemistry	Water chemistry
	Mineralogy, petrology, and rock chemistry
	Stability of minerals and glasses
	Radionuclide retardation by sorption
	Radionuclide retardation by precipitation
	Radionuclide retardation by dispersive, diffusive, and advective processes
	Radionuclide retardation by all processes
	Retardation of gaseous radionuclides
Rock characteristics	Strategy for integrated drilling program
	Geologic framework of the site
	Three-dimensional models of rock characteristics
Climate	Rates of change in climate
	Effects of future climate on hydrologic characteristics
Erosion	Locations and rates of surface erosion
	Effects of future climate on locations and rates of erosion
	Effects of future tectonic activity on locations and rates of erosion
Postclosure tectonics	Volcanic activity
	Waste-package failure due to tectonic events
	Hydrologic changes due to tectonic events
	Changes induced by tectonic processes in the geochemical properties of the rocks
	Data collection
Human interference	Activities that might affect surface markers and monuments
	Value of natural resources
	Effects of exploiting natural resources
Meteorology	Regional meteorological conditions
	Local meteorological conditions
	Atmospheric and meteorological phenomena at the site
	Population centers relative to wind patterns
	Extreme-weather phenomena
Offsite installations	Determination of nearby industrial, transportation, and military installations and operations
	Potential impacts of nearby installations and operations
Surface characteristics	Topography of potential locations for surface facilities
	Soil and bedrock properties

Table II.—Continued

Characterization Program	Investigation
Thermal and mechanical rock properties	Spatial distribution of thermal and mechanical properties
	Spatial distribution of ambient stress and thermal conditions
Preclosure hydrology	Flood recurrence intervals and levels
	Locations of adequate water supplies
	Groundwater conditions within and above the potential host rock
Preclosure tectonics	Volcanic activity
	Fault displacement
	Vibratory ground motion
	Preclosure-tectonics data collection and analysis

NOTE: This table was created in December 1988.
SOURCE: U.S. Department of Energy. *Investigations To Be Conducted in the Site Program*; DOE/RW-0198; Government Printing Office: Washington, DC, December 1988.

process that resulted in the initiation of studies at Yucca Mountain and to have a general description of the site.

Site Screening

The process that led to the Yucca Mountain studies began in 1977, when the U.S. government decided to investigate the possibility of siting a repository at the Nevada test site (NTS). The NTS, about the size of Rhode Island and located in southern Nevada, was selected for this investigation because nuclear activities were familiar to the area. The land that the NTS occupies is federally owned and was withdrawn from public use for nuclear defense testing activities. In addition, the U.S. Geological Survey (USGS) proposed that the NTS be considered for siting a geologic repository for spent fuel and high-level radioactive waste for a number of geologic reasons, including the following:

1. In southern Nevada groundwater does not discharge into rivers that flow to major bodies of surface water—the most likely pathway for radionuclides to reach people and the environment.

2. The kind of rock formations found at the NTS have geochemical characteristics that are favorable for waste isolation: they would retard the migration of radionuclides.

3. The paths of groundwater flow between potential sites for a repository and the points of groundwater discharge

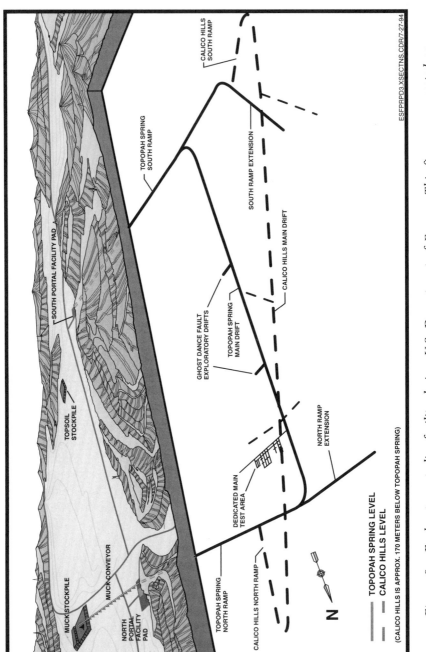

Figure 6. *Exploratory studies facility design. U.S. Department of Energy. This figure was created on July 27, 1994.*

are long: the water table at the Yucca Mountain area is about 2300 ft (690 m) below the crest of Yucca Mountain.

4. Because the region is arid the rate at which groundwater is recharged is very low, and therefore the amount of moving groundwater is also very low, especially in the unsaturated rocks.

So that geologic repository studies would not interfere with weapons-testing activities at the NTS and vice versa should Yucca Mountain ever become a repository, site screening was eventually limited to the southwestern part of the NTS and adjacent land that is also federally owned. (The adjacent land is owned by the U.S. Bureau of Land Management, and parts of that land are leased to the Nellis Air Force Base.)

One of the three locations identified on this land was Yucca Mountain. Yucca Mountain contains a large block of volcanically formed rock called "tuff". That block seemed to be large and thick enough to be considered for a repository. Previously, in other parts of the country, salt and basalt were the geologic media studied as potential host rocks. Since tuff had not previously been considered as a potential host rock for a repository, the federal government sought the views of the NAS. The NAS responded favorably.

At the time of the NAS study, the USGS recommended Yucca Mountain be studied as a possible repository candidate after comparing the results of preliminary investigations at all three locations in the area.

In 1980 a formal federally directed scientific analysis of 15 potential locations showed that Yucca Mountain was the preferred candidate site. In February 1983, following passage of the NWPA, Yucca Mountain was formally identified as one of nine potentially acceptable sites for a repository.

In December 1984 the DOE issued a draft environmental assessment on Yucca Mountain (as well as on each of the other eight potentially acceptable sites) and announced its proposal to nominate the Yucca Mountain site as one of five sites suitable for characterization and to recommend Yucca Mountain and two other sites as candidates for a repository.

Following public meetings and public hearings on these environmental assessments and the proposed nominations and recommendations, in May 1986, the Secretary of Energy issued final environmental assessments on Yucca Mountain and the other potential sites and formally recommended to President Reagan that Yucca Mountain and two other sites (Hanford in the State of Washington and Deaf Smith in Texas) be characterized as candidates for a repository. The president approved the recommendation.

Following this action numerous lawsuits were filed by the states selected. Some of these lawsuits regarded the process of selection. Subsequently, extensive debate occurred within Congress and resulted in the introduction of 40 bills to change the national program for developing the waste disposal system. Some of these bills recommended the program speed up, some wanted it to slow down, and others hoped it would cease and start over. Some bills wanted the government to look at more potential sites, and some wanted them to study only one site at a time.

On December 21, 1987, as a result of these public and political actions, Congress acted again on the possible repository-siting issue. Almost 5 years after the passage of the original act, Congress passed the Nuclear Waste Policy Amendments Act of 1987 (12), which, among other things, directed the DOE to characterize only the Yucca Mountain site as a candidate for the first repository. The Amendments Act further directed the DOE to terminate siting activities at all other sites.

Description of Yucca Mountain

Yucca Mountain is in southern Nevada in Nye County, about 100 miles northwest of Las Vegas (Figure 7). It is on three parcels of federal land (Figure 8) and lies in the southern part of the Great Basin, an arid region with linear mountain ranges and intervening valleys. The Great Basin receives very little rainfall (about 6 in. (15.2 cm)/year), and has both sparse vegetation and a sparse population.

At the crest of Yucca Mountain the elevation is about 5000 ft (1500 m) above sea level. The water table at Yucca Mountain is very deep, about 2300–2500 ft (690–750 m) below the surface. Because the rainfall is so low and evaporation of the rainfall is so high, there is little percolation of water downward through the unsaturated rock, the possible location for a repository, about 1000–1200 ft (300–360 m) below the surface and about 700–1000 ft (210–300 m) above the water table.

The origin of the tuff rock found in the Yucca Mountain region is from volcanic eruptions that occurred 8 to 16 million years ago. The thickness of the volcanic rock that formed from these eruptions is about 6500 ft (1950 m). The molten materials explosively expanded upon eruption and broke into particles of hot glass. These particles spread across the surrounding land and eventually came to rest. The particles were then subjected to various degrees of compaction and fusion, depending on temperature and pressure. In areas where the tempera-

Figure 7. Map of the United States with Yucca Mountain photograph projected. Yucca Mountain has been named by Congress for site characterization in the United States. U.S. Department of Energy. This figure was created in August 1991.

ture and pressure were high enough, a rock known as "welded tuff" formed.

To judge whether Yucca Mountain is geologically suitable for a geologic repository for spent fuel and high-level radioactive waste, all significant processes and events important to waste isolation must be considered. These processes and events include the natural ones that are expected to occur at the site over the next 10,000 years and the potentially disruptive ones that are not expected but are sufficiently credible to warrant consideration. Geologic history of approximately 2 million years (the Quaternary period in geologic time) is important in making judgments about future performance of the site.

The geologic history of Yucca Mountain suggests that the phenomena of special interest are the effects of faulting, seismicity, and volcanic activity. Understanding past faulting and volcanic activity will provide the basis for determining the potential for disruptive tectonic events during both the operation (when spent fuel and high-level ra-

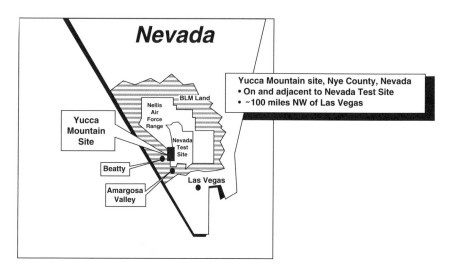

Figure 8. Map of Nevada showing use of federally owned land. U.S. Department of Energy. This figure was created in July 1991.

dioactive waste are being received) and the closure of a repository. Information on seismicity at and around a potential site is an important element in determining the design of the surface facilities of a repository. Many other areas also are important to assessing the suitability of the site, including the occurrence of natural resources. The exploration for such resources could lead to future intentional or unintentional human intrusion into a repository.

Although studies of Yucca Mountain as a possible geologic repository for spent fuel and high-level radioactive waste did not begin until 1977, information about the geologic history and conditions around Yucca Mountain was collected dating back to the early 1900s. Information was first collected to support exploration for mineral and energy resources and later to support federal activities at the NTS.

But only since 1977 was information collected specifically relative to siting a geologic repository for spent fuel and high-level radioactive waste. This information was obtained by reviewing published data, performing detailed geologic mapping of the Yucca Mountain area, conducting regional geophysical investigations, recording seismic-monitoring data, and conducting other field studies.

Volcanism, Faulting, and Seismicity

Yucca Mountain, as mentioned earlier, was formed as a result of a volcanic eruption 8 to 16 million years ago; however, there are several other volcanic features in the vicinity of Yucca Mountain that possibly formed only 10,000 to 20,000 years ago. Volcanic activity in the area is well documented. The data contained in this documentation suggest that the probability of major, disruptive volcanic activity occurring at Yucca Mountain is negligible. On the other hand, there may be a higher possibility that volcanic activity characterized by minor, low-volume disruptions of short duration may occur over the next 10,000 years.

The geologic structure of Yucca Mountain is extremely complex and contains a variety of faulting, which mainly occurred in response to the tectonic activity associated with volcanism 7–11 million years ago. Data on faulting are more important to the design of surface facilities than for underground facilities, and, in this regard, data will be collected and evaluated concerning movement during the Quaternary Period, about 2 million years ago. Evidence of this movement can be seen at the surface and in shallow trenches.

During the period of time that records are available, the past 150 years, eight major earthquakes (6.5 or more on the Richter scale) have occurred within about 250 miles (402.5 km) of Yucca Mountain, which is located about 100 miles (161 km) east of the Nevada–California seismic belt and about 150 miles (241.5 km) northwest of the Intermountain seismic belt. Six of these earthquakes occurred in the Nevada–California seismic belt, and two occurred on or near the San Andreas fault. Until 1992 the nearest recorded major earthquake was the 1972 Owens Valley earthquake about 90 miles (144.9 km) west of Yucca Mountain. This earthquake had a magnitude of about 8.3.

Geologic field evidence suggests that, in terms of major tectonic activity, Yucca Mountain has been relatively stable for the past 11 million years. Recent seismic data are available from a 47-station seismic network that was installed within 100 miles (161 km) of the site in 1978 and 1979 and a supplemental six-station network that was installed at Yucca Mountain in 1981. Measurements made since 1978 show that, within about 6 miles (9.7 km) from the Yucca Mountain study area, the release of seismic energy has been 100–1000 times lower than that in the surrounding region.

Preliminary designs developed in the mid-1980s and contained in the SCP for Yucca Mountain were based on possible earthquakes with a magnitude of 6.8. Based on more recent occurrences and data from the 1990 San Francisco earthquake and earthquakes in June 1992 in California and at Little Skull Mountain, approximately 40 miles (64.4 km) from the Yucca Mountain study site, additional seismic monitoring stations were installed near Yucca Mountain and increased data collection was initiated. The California earthquake which occurred on

June 28, 1992, measured about 6 on the Richter Scale, and the Little Skull Mountain earthquake measured about 5.6.

The June 1992 earthquake at Little Skull Mountain caused some damage to a temporary building located near Yucca Mountain. Some windows were broken and some ceiling tiles fell. The structure where damage occurred was a temporary building: it was neither designed nor built to withstand surface disturbances. A repository and the surface structures associated with the repository can be, will be, and were always planned to be designed and built to withstand major earthquakes. The 1992 earthquakes provided valuable opportunities for scientists to collect data and perform analysis of the area.

Geoengineering

The behavior of the rock, or tuff, as an engineering material must be understood in order to design, construct, operate, and close a repository. While a repository is, or will be, a "mine", it is different from ordinary mines and tunnels for a few reasons, as follows: (1) the spent fuel and high-level radioactive waste to be emplaced in it will add heat and radiation to the rock mass; (2) a need for long-term stability of the rock is necessary; and (3) because of the size and nature of the rock mass required for a repository, roof bolts and wire mesh should be sufficient to stabilize the openings and no unusual support systems should be required during the excavations for exploratory work or for a repository.

The heat from the radioactive waste will change the temperature field of the repository, which, in turn, will change the state of stress and possibly the distribution and flow of moisture in the rock mass. Geoengineering properties are important in the construction and operation of a repository, because they control the stability of the waste-emplacement holes. The latter affects worker safety, waste retrievability, and the integrity of the waste container.

The database for the geoengineering properties of the tuff at Yucca Mountain is derived from two sources: (1) the results of laboratory tests on small-diameter (2.5-in. or 6.4-cm) core samples from Yucca Mountain and outcrop samples in the vicinity of the site; and (2) both field and laboratory tests on similar tuff units in the region.

Hydrology

The hydrologic conditions are critical to long-term performance of a repository. They may affect the behavior of the waste package, as the movement of groundwater is the principal mechanism for transporting radionuclides to the accessible environment.

An important and attractive feature of the geohydrology at Yucca Mountain is that it is located in the unsaturated zone. The unsaturated zone is the rock mass between the surface of the land and the water table. At Yucca Mountain the unsaturated zone is believed to be thick enough to allow the construction of a repository about 700–1300 ft (210–390 m) above the top of the water table. Beginning in the early 1980s, test holes deeper than 1000 ft (300 m) were drilled into the unsaturated zone and used to monitor the ambient water saturation, potential, and flux in the rocks above, below, and at the depth of a proposed repository.

At Yucca Mountain, similar to the rest of the Southern Great Basin within which Yucca Mountain is located, the groundwater basins tend to be closed, with no external drainage into rivers or major bodies of surface water. Furthermore, current estimates of groundwater travel time from the proposed repository to the underlying water table range from about 9,000–80,000 years. Additionally, no perennial streams occur at or near Yucca Mountain.

Geochemistry

The geochemical environment of the host rock may also affect the long-term performance of a repository by affecting the behavior of the waste package and by retarding the transport of radionuclides. Geochemical data were collected since the late 1970s from rock samples taken at or in the vicinity of Yucca Mountain. Samples for mineralogic and petrologic studies were taken from drill cores, drill cuttings, and surface outcroppings. Data on water chemistry have been obtained from groundwater samples taken from deep wells, and information on the stability of geochemical conditions was obtained from laboratory experiments.

An attractive geochemical feature of the Yucca Mountain area is the existence of minerals with a high sorption capacity that are present along potential paths of groundwater flow below the repository and in the saturated zone below the potential repository depth. One of these features is zeolites.

Sorption is a process for removing dissolved material from solution by attaching the dissolved solids to the surface of another solid. Sorptive capacity of a rock is a measure of the ability of its surfaces to remove dissolved material from solutions passing through the rock. The sorptive capacity of the rock will work to limit waste movement.

Sorptive capacity includes both chemical and physical processes. The chemical processes include all those mechanisms that can take atoms or molecules and attach them to rock (or mineral) surfaces. An example of a chemical process of sorption is ion exchange. In an ion exchange, particles with an electrical charge (ions) in water change places with ions that are attached to a mineral, such as zeolites. In the case of a potential repository where zeolites are present, the radioactive ions are removed from the water, and attached to the zeolites, leaving only the harmless ions to go into the water.

A zeolite is a naturally occurring ion exchanger. Ion exchange is based on the simple idea that electrically charged particles with unlike charges can attract and neutralize each other. An ion is an atom, molecule, or molecular fragment carrying a positive or negative electrical charge. Ion exchange is the process in which an ion in solution takes the place of another ion in a natural or man-made material. In other words, if rocks surrounding a repository have ion-exchange capability, which the existence of zeolites provides, they present a natural barrier to the movement of radionuclides by "exchanging" harmless ions within their structure for the radionuclides moving past them. Because the zeolites are positioned above the water table, radioactive elements coming from the repository can be filtered, preventing or delaying the migration of radionuclides. The existence of zeolites is significant because the natural process that occurs if zeolites are present in the rock surrounding a proposed repository can help protect the environment and people by removing contaminants from water.

Climatology and Meteorology

The existing climate in the vicinity of Yucca Mountain is classified as a midlatitude-desert climate. The general meteorological characteristics of such a climate are temperature extremes, particularly during the summer months, approaching 49 °C; large ranges in the high and low temperatures; and an annual precipitation of less than 6 in. (15.2 cm)/year.

Climatic changes that may occur in the next 10,000 years are important to the long-term performance of a repository, because a change from the current arid conditions at Yucca Mountain might affect hydrologic conditions. At Yucca Mountain the potential for a change in the amount of groundwater flux through the unsaturated zone and a rise in the water table are important, because the thickness of the unsaturated zone below the repository could be decreased and the amount of water available for contact with the waste containers or the waste itself could be increased.

As with other potential future physical events, climatic trends that are expected or predicted into the next 10,000 or more years will be based on changes and trends that occurred over the past 2 million years.

Methods used for determining past climates and climatic changes include the study of plant remains left thousands of years ago in the middens of pack rats, fossilized plant pollens, evidence of past lake positions preserved in marshes and deposits formed along what were shorelines, and any glaciation that may have occurred in the area.

Technological Advances

In the process of developing this first-of-a-kind system, technological advances are expected. Some advances already occurred and are commercially used. One prominent advance is the design and fabrication of a new deep, dry drill rig. Most deep drilling and coring techniques require the use of water as a lubricant and as a drill-bit cooler in the operation of the equipment. To understand the geohydrologic characteristics of the tuff at Yucca Mountain, core must be obtained without introducing water and without changing the geohydrological nature of the rock.

The LM-300 dry drill rig was designed and built for the DOE by Lang Drilling Company in Utah. It is capable of drilling dry and extracting core some 3000 ft (900 m) deep. After extensive prototype testing and training of personnel, the DOE transported the LM-300 to Nevada and began drilling in late May 1992. During the first 4 months of operation, the LM-300 successfully reached a depth of approximately 900 ft (270 m). At depths of 1000 ft (300 m), core representative of the potential repository-level rock was obtained. Beyond that depth, up to 3000 ft (900 m) deep, core was obtained to evaluate the rock characteristics below the potential repository.

External Oversight and Societal Aspects

The NWPA, as amended (13), made the DOE responsible for developing and operating the system to provide safe and environmentally acceptable storage, transportation, and permanent disposal of spent fuel and high-level radioactive waste. The law also provides for extensive oversight and independent review by special federal bodies, state and local governments, and the public.

Like other countries developing such waste management and disposal systems, the U.S. DOE invites scientific and public oversight and review. The difference between the U.S. program and the programs of other countries, however, is that the United States actually contributes funds for the oversight and review, adding cost and time to interact with the overseers and reviewers.

Since the identification of Yucca Mountain in early 1983 as a potentially acceptable site for a geologic repository, the DOE has provided grant funds from the Nuclear Waste Fund for the State of Nevada to participate in and oversee the national program. However, until a U.S. Supreme Court decision in 1991 required the state to process the environmental permit requests required before the DOE could proceed with necessary site characterization studies from 1986 to 1991, the DOE was prevented by the State of Nevada from conducting new on-site scientific investigations.

The law also provides that affected local governments or Indian tribes also receive grant funds to participate in and oversee the program. As of early 1993 nine Nevada jurisdictions and one California jurisdiction were designated as affected and received grant funds from the Nuclear Waste Fund. No Indian tribes had been designated as affected as of that date.

In 1987 the Nuclear Waste Policy Amendments Act authorized the establishment of the MRS Review Commission, the Nuclear Waste Technical Review Board (NWTRB), repository and MRS review panels, and the Office of the Nuclear Waste Negotiator—all funded from the Nuclear Waste Fund.

The MRS Review Commission was charged with preparing a report to Congress on the need for an MRS facility as part of a waste-management system. This was, in effect, a second opinion, since the Amendments Act had already authorized an MRS. Following extensive public meetings and interactions with the DOE, the MRS Review Commission submitted its report to Congress, agreeing that a temporary or interim facility, such as an MRS facility, would be beneficial to the overall waste-management and disposal system.

The NWTRB is an independent board that contains 11 members appointed by the President from candidates nominated by the U.S. National Academy of Sciences. The NWTRB is authorized a staff, charged with evaluating the technical and scientific validity of activities of the waste-management program, and must report to Congress and the Secretary of Energy at least twice a year. The NWTRB will exist until a year after the repository begins operations. Because 2010 is the earliest a repository might begin operation (if Yucca Mountain is found suitable), the NWTRB will exist until at least 2011.

Many other panels were established by the Amendments Act and funded by the Nuclear Waste Fund to review or conduct special studies concerning the waste management and disposal system. Additionally, a variety of cooperative agreements between the DOE and national or regional organizations are funded by the Nuclear Waste Fund to participate in the program.

The Amendment Act requires extensive general public involvement in the program through document review and public meetings and hearings. The program also receives much attention from the media.

Even though U.S. policy has been established, the debate continues over the solution to the nuclear-waste-management problem. Opposition or concern by individuals and public interest groups regarding possible radiation exposure at or near proposed facilities and as a result of possible transportation accidents continues. Many Americans are concerned about protecting the environment and preventing unnecessary disturbances in the future.

The nuclear-waste-management program in the United States and in other countries is an evolving program with many decisions still to be made: scientific, technological, and societal. Many questions remain.

Increasingly, decisions that effect the long-term well-being of society and the environment involve complex scientific and technological issues. The wise use of technology in solving scientific and societal problems is essential. Making decisions about energy and the environment and successfully protecting both public health and safety and the environment relative to nuclear-waste management require the extensive integration of science and society.

References

1. "Nuclear Waste Policy Act of 1982"; Public Law 97-425, 1983.
2. "Atomic Energy Act of 1954"; Public Law 703, 1954.
3. MRS Review Commission. *Nuclear Waste: Is There a Need for Federal Interim Storage?*; U.S. Government Printing Office: Washington, DC, 1989.
4. "Low-Level Radioactive Waste Policy Act of 1980"; Public Law 99-240, 1980.
5. U.S. Department of Energy. *Integrated Data Base for 1993: U.S. Spent Fuel and Radioactive Waste Inventories, Projections, and Characteristics*; DOE/RW-0006, Rev. 9; Washington, DC, March 1994.
6. U.S. Department of Energy. *Environmental Assessment, Hanford Site*; DOE/RW-0070; Reference Repository Location, Washington, DC, May 1986; Vols. 1–3.
7. U.S. Department of Energy. *Environmental Assessment, Yucca Mountain Site*; DOE/RW-0073; Nevada Research and Development Area, Nevada, May 1986; Vols. 1–3.
8. U.S. Department of Energy. *Site Characterization Plan, Yucca Mountain Site*; DOE/RW-0199; Nevada Research and Development Area, Nevada, December 1988; Vols. 1–9.
9. U.S. Department of Energy. *Environmental Assessment, Richton Dome Site*; DOE/RW-0072; Mississippi, May 1986; Vols. 1–3.
10. U.S. Department of Energy. *Environmental Assessment, Deaf Smith County Site*; DOE/RW-0069; Texas, May 1986; Vols. 1–3.
11. U.S. Department of Energy. *Environmental Assessment, Davis Canyon Site*; DOE/RW-0071; Utah, May 1986; Vols. 1–3.

12. "Nuclear Waste Policy Amendments Act of 1987"; Public Law 100-203, 1987.
13. "Nuclear Waste Policy Amendments Act of 1988"; Public Law 100-507, 1988.

RECEIVED for review October 2, 1992. ACCEPTED revised manuscript April 29, 1993.

GLOSSARY

ABCC	Atomic Bomb Casualty Commission.
Abscissa	The coordinate representing the distance of a point from the vertical y-axis parallel to the horizontal x-axis.
ABWR	Advanced boiling water reactor.
ACNS	Advisory Committee on Nuclear Safeguards.
Actinides	A series of elements beginning with actinium, atomic number 89, through lawrencium, atomic number 103; the 5f electronic shell is filled in this series.
Acute exposure	The absorption of a relatively large amount of radiation in a short period of time.
AEC	Atomic Energy Commission.
AECL	Atomic Energy of Canada Limited.
AEDE	Annual effective dose equivalent.
ALARA	As low as reasonably achievable.
Aleukemic	An absence of white cells in blood. Leukemia without the presence of typical leukemic white cells.
ALI	Annual limit of intake; refers to the occupational-effective-dose-equivalent-limit of a radionuclide taken into the body in 1 year.
Alpha particle	The nucleus of a helium atom ejected from some radionuclides when they decay.
AMAD	Activity median aerodynamic diameter.
Aneuploids	Cells that have numbers of chromosomes not equal to multiples of the haploid number; cells that have an unbalanced set of chromosomes.

0065–2393/95/0243–0315$08.00/0

Ankylosing spondylitis	Arthritis of the spine.
ANL–ACRH	Argonne National Laboratory–American Cancer Research Hospital.
APWR	Advanced pressurized water reactor.
ATB	At time of bombing.
Basal cell	The target cell at risk in lung cancer. These cells are the dividing stem cells in the bronchial epithelium.
BEIR	Biological effects of ionizing radiation.
Beta particle	An electron or positron (positive or negative charge) emitted by an atomic nucleus or neutron in a radioactive decay process.
Biokinetic	Rate of uptake, retention, and excretion of a radionuclide in various organs of the body.
Body burden	The quantity of radioactive material in an individual's body at a particular time.
Bone seeker	Any compound or ion that preferentially migrates into bone.
Bq	Becquerel, the SI unit of activity; 1 Bq is one nuclear disintegration/second.
Brachytherapy	A method of radiation therapy in which an encapsulated radioactive source is placed or implanted within, or close to, the area to be irradiated.
Cancer	A malignant new growth of tissue that invades the body and tends to spread.
CANDU	Canadian deuterium uranium (reactor).
Carcinogen	An agent, chemical or physical, that causes cancer.
Carcinoma	A malignant tumor.
CBD	Chronic beryllium disease.
CDC	Centers for Disease Control and Prevention.
CEDE	Committed effective dose equivalent.
CEDR	Comprehensive epidemiologic data resource.

Cell	The smallest structural unit of an organism that can function independently. It can consist of a nucleus, cytoplasm, and various specialized parts surrounded by a semipermeable membrane.
Cell death	Loss of a cell's structural integrity, reproductive ability, or functional activity.
Cell survival	Ability of a cell to proliferate.
Cells, somatic	Body cells.
CERCLA	Comprehensive Environmental Response, Compensation and Liability Act (Superfund).
Chernobyl	A nuclear reactor located in Chernobyl, Ukraine, near Kiev.
CHR	Center for Human Radiobiology.
Chromosome	A linear body in a cell nucleus that contains genes, the hereditary information of the cell.
Chronic	Continuous over a long period of time.
Ci	Curie, a unit of radiation; *see* Curie.
Cohort	A large homogeneous group of people; an epidemiological term.
Collective dose	The integral dose to a group of people.
Committed dose equivalent	The total equivalent radiation dose to the body, or specified part, accumulated over 50 years after intake of a source.
Committed effective dose equivalent	The weighted sum of committed dose equivalents to individual organs and tissues.
Confidence interval	The measure of the reliability of a risk estimate; a statistical reliability probability.
Confounding factor	Uncontrollable variable that may or may not be important in an epidemiological study.
Cosmic rays	Ionizing radiation of extraterrestrial origin, which can be protons, helium, other atomic nuclei, high-energy electrons, or photons.
Critical organ	That organ in which a particular radiation pattern dose would be most significant.
Cumulative dose	Total dose resulting from repeated radiation exposure to a defined region.

Curie

A unit of radioactivity equal to 3.7×10^{10} nuclear transformations per second.

Cytogenetics

A study of heredity by cytological and genetic techniques.

Cytogenic

Pertaining to cytogenetics.

DAC

Derived air concentration; the concentration of a radionuclide in air that if breathed by a reference man for 1 year would yield an ALI.

Daughter

The atomic species that is the product element of a radioactive event.

De novo

Anew; fresh; a new occurrence.

Decay

Disintegration of the nucleus of an unstable nuclide.

Delta ray

A secondary ionizing particle ejected by recoil when an energetic particle passes through matter.

Depth-dose curve

A profile of the absorbed radiation dose as a function of depth into a material.

Deterministic model

A model whose output is predetermined by mathematical models and selected single value input parameters.

Deuterium

An isotope of hydrogen containing one proton and one neutron in the nucleus.

Directly ionizing radiation

A radiation composed of electrically charged particles capable of ionizing other matter.

Distribution factor

A factor used for dose equivalent; this factor allows for nonuniform internal distribution of radioactive nuclides.

DNA

Deoxyribonucleic acid.

DOE

Department of Energy.

Dose

Quantity of radiation energy absorbed, generally per unit mass.

Dose, absorbed

The energy imparted by ionizing radiation to a volume element of matter divided by the mass of irradiated material in the volume element. The unit of absorbed dose is the gray (Gy) or rad; 1 Gy = 100 rad.

Dose, median lethal	An absorbed dose that will kill 50% of a species in a defined time period, also known as LD_{50}.
Dose, occupational	The radiation dose received at the workplace.
Dose, skin	The absorbed dose on an area of skin.
Dose, threshold	Minimum dose that will produce a particular effect.
Dose commitment	The total dose-equivalent received by a body, or part thereof, over a hypothetical 50-year period following some initial intake.
Dose equivalent	The product of absorbed dose, quality factors, and other modifying elements in order to normalize the dose to a common scale; units of dose equivalent are the rem or sievert.
Doubling dose	The amount of radiation required to double the natural incidence of some genetic anomaly.
DS 86	Dosimetry system 1986 (*see* the chapter by Yoshimoto).
Dycrasia	A nonspecific pathologic condition usually referring to cellular components.
Effective dose equivalent	The sum of the weighted average of organ dose equivalents.
Effective half-life	The time required in a biological system for a given radionuclide's activity to be halved; this reduction takes into account radioactive decay and biological elimination.
Electrophoretic	Having to do with the motion of charged particles or ions in a stationary liquid phase by the application of an electric field.
EPA	Environmental Protection Agency.
Epidemiology	The study of health and sickness in divisions of human populations.
Epidermis	The outermost layer of skin.
Epilation	Loss of hair.
Epithelium	The membrane that forms the covering of organs and the outer surface of an animal body, usually a single layer of cells.
EPRI	Electric Power Research Institute.

ERDA	Energy Research and Development Administration.
Erythrocyte	Red blood cell.
Exposure, acute	*See* Acute exposure.
Exposure, chronic	Radiation exposure of prolonged duration.
Fallout	Radioactive debris deposited from airborne particulates.
FDA	Food and Drug Administration.
Fission	The division of a heavy nucleus into two (or rarely more) nuclear fragments; neutrons are usually released during this type of transmutation.
Free radical	An atom or molecule having at least one unpaired electron.
Gaia	The earth.
Gamma ray	High-energy electromagnetic radiation of nuclear origin.
General public	Mass of population not regarded as radiation workers.
Genetic effect	A change produced in that part of a cell that controls heredity; it can be caused by radiation.
Genetics	That branch of biology dealing with heredity.
Granulocyte	A leukocyte whose cytoplasm contains stainable granules.
Gray	The SI unit of absorbed radiation dose; 1 Gy = 1 J/kg = 100 rad.
GWe	Gigawatts, electrical; 10^9 W.
Gy	Abbreviation for gray, a unit of absorbed dose.
Half-life, biological	Time required for the body to eliminate one-half of an introduced substance.
Half-life, effective	*See* Effective half-life.
Half-life, radioactive	The time required for the activity of some radioactive material to be reduced to one-half its original value.

Half-value layer	Thickness of an absorber that reduces the intensity of a radiation beam by one-half.
Hard radiation	Radiation that has high penetrating character.
Health physics	The science concerned with radiation protection.
Healthy worker effect	The fact that the working population is healthier than the average general population.
HEHF	Hanford Environmental Health Foundation.
Hematocrit	The percent of red blood cells in a volume of blood.
Hematopoietic	Blood forming.
Hemoglobin	The oxygen-bearing protein in red blood cells.
HHS	Department of Health and Human Services.
HMS	Health and Mortality Study.
Hormesis	The theory that suggests that a little radiation is beneficial to living organisms.
HTGR	High-temperature gas-cooled reactor.
Hypertension	High arterial blood pressure.
Hyperthyroidism	A disease state characterized by an excess of thyroid hormones in the blood.
IAEA	International Atomic Energy Agency.
IARC	International Agency for Research on Cancer, World Health Organization.
ICP	International Chernobyl Project.
ICRP	International Commission on Radiological Protection.
ICRU	International Commission on Radiation Units and Measurements.
INEL	Idaho National Engineering Laboratory.
Internal dosimetry	The measurement or estimate of internal radiation dose.
Internal emitter	A radionuclide deposited in the body.
Internal radiation	Radiation emitted by a radionuclide within the body.
Ionizing radiation	Radiation consisting of charged particles or particles that can ionize other matter.

Isotope	Name given to nuclides of a given element that have a unique number of neutrons.
ITRI	Inhalation Toxicology Research Institute.
Kerma	The measure of initial kinetic energy of charged particles that are released from the interaction of uncharged particles, such as neutrons or X-rays, with matter. Units of kerma are grays.
kg	Kilogram, 1000 g.
kGy	Kilogray, 1000 Gy.
Koseki	Official Japanese family registries.
kW	Kilowatt, 1000 W.
LANL	Los Alamos National Laboratory.
Latent period	Period of time between exposure to an agent and the start of a measurable response to that agent.
Lesion	A defined or circumscribed area on or within the body that has been injured or is the site of a disease process.
LET	Linear energy transfer.
Leukemia	A cancerous blood disease in which there is an excessive production of white blood cells.
Leukemogenic	Causing leukemia.
Leukocyte	White blood corpuscle.
Linear hypothesis	Hypothesis that risk is proportional to dose.
Linear model	Effect is proportional to dose.
Linear-quadratic model	Effect has both linear and higher ordered proportionality to dose, the former at low doses and the latter at high doses.
LMR	Liquid metal reactor.
LSS	Life span study.
LWR	Light water reactor.
Lymphocytic	Pertaining to a variety of white blood cells formed in lymphoid tissue.
Lymphoma	Cancer of lymphoid tissue.
Malignancy	Abnormal growth that can metastasize; cancer.

Man-rem	The unit of population exposure in rems that is the sum of all individual dose equivalents for all members of that population.
Mandibular osteomyelitis	Infection of the bone marrow of the jaw.
Manhattan Project	The name given to the secret project that developed the atomic bomb.
MBq	Million becquerels, 10^6 Bq.
MED	Manhattan Engineering District.
Melanoma	A nonsymmetric, dark cancerous tumor.
Metastasis	Transfer of cancerous cells from a primary site to other locations.
MeV	Million electron volts.
MHTGR	Modular high-temperature gas reactor.
Microorganism	A microscopic plant or animal.
Mill tailings	Sediment from a mining operation.
Monazite	A phosphate mineral that contains rare earths and sometimes uranium and thorium.
Morbidity	The state of being diseased.
Mortality	The frequency of death in a segment of a population.
MOU	Memorandum of understanding.
MPC	Maximum permissible concentration.
MRS	Monitored retrievable storage.
mSv	Millisievert.
Mutagen	An agent that causes a heritable change in the genes of an organism.
Mutant	An organism that has undergone a mutation.
Mutation	A heritable change in the genes of an organism.
MWe	Megawatts, electrical, 10^6 watts electrical.
MWt	Megawatts, thermal, 10^6 watts thermal.
Myelogenous	Produced in the bone marrow.
Myeloma	A malignant tumor composed of cells normally found in bone marrow.
NAS	National Academy of Science.

NCI	National Cancer Institute.
NCRP	National Council on Radiation Protection and Measurements.
Necrosis	The death of living tissue.
Neoplasm	A new growth of cells not restrained by normal reproductive processes; a tumor, either benign or malignant.
NERVA	Nuclear engine for rocket vehicle application.
Neutron	An uncharged elementary particle that is normally located in an atomic nucleus; it is similar in mass to the proton.
NIOSH	National Institute of Occupational Safety and Health.
NIST	National Institute of Standards and Technology.
NJRRP	New Jersey Radium Research Project.
Nonstochastic	Not random; describes effects whose severity is proportional to dose.
NRC	National Research Council *or* Nuclear Regulatory Commission.
NTS	Nevada test site.
Nuclear	Related to the nucleus of an atom.
Nuclear fuel	The fuel for a nuclear reactor.
Nuclear medicine physician	A physician who deals with diagnostic and therapeutic applications of radionuclides.
Obesity	The condition of being overweight.
Occupational exposure	The exposure of an individual to ionizing radiation in the course of employment.
OEHS	Office of Epidemiology and Health Surveillance.
ORAU	Oak Ridge Associated Universities.
Organ burden	The quantity of a radionuclide present in an organ of the body at a particular time.
Organ weighting factor	Relative risk of inducing cancer by radiation of a given organ or tissue.
ORNL	Oak Ridge National Laboratory.

Osteitis	Inflammation of bony tissue.
PAREP	Populations at risk from environmental pollutants.
Pathogen	A disease-causing agent.
Penetrating radiation	External radiations of sufficient penetrating power to expose tissues beneath the skin; examples are gamma, X-ray, and neutron radiation.
Person-gray	Unit of population exposure, expressed in grays, representing the sum of the dose equivalent values for all people in an exposed population.
Person-rem	Same as person-gray except expressed in rems.
PIUS	Process inherent ultimately safe.
Plutonium	Element with atomic number 94, a man-made, radioactive element.
PNL	Pacific Northwest Laboratory.
Poisson	Name given to a probability distribution in which the mean and the variance (the square of the standard deviation) of a measurement are equal. It describes the results of counting a radioactive sample.
Prescience	Knowledge of events before they occur.
Protein	Various groups of nitrogenous organic compounds that contain amino acids; these compounds are essential to life.
Proton	An elementary positively charged particle generally found in the nucleus of an atom. The nucleus of the common isotope of hydrogen.
^{239}Pu	A radioactive isotope of plutonium that has an atomic weight of 239.
Pyrolytic carbon	A form of carbon.
Quality factor	The multiplier used with absorbed radiation dose to define its relative effectiveness.
Quintile	The portion of a statistical distribution containing one-fifth of the population.
Ra	Symbol for the element radium, atomic number 88.

Rad	The unit of absorbed radiation dose equal to 100 ergs/g; 1 rad = 0.01 Gy.
Radiation	Energy propagated through space.
Radiation, external	Radiation from a source outside the body.
Radiation, internal	Radiation from a source inside the body.
Radiation, ionizing	Radiation capable of producing ions as it travels through matter.
Radioactive decay	Spontaneous transformation of a nucleus with the emission of a particle or particles, and/or photon; rate of decay is first order.
Radioactivity	A general term relating to the emissions observed in radioactive decay.
Radiocesium	One or more radioactive isotopes of the element cesium, usually ^{134}Cs or ^{137}Cs.
Radioimmuno-assay	A method for measuring small quantities of biological substances, such as hormones, that uses an antibody labeled with a radioactive tracer to react with the substance to be measured.
Radiologist	A physician who deals with diagnostic and therapeutic applications of X-rays.
Radionuclide	Any radioactive form of an element.
Radiosensitivity	The susceptibility of cells, tissues, organs, etc., to damage by ionizing radiation.
Radium	Element of atomic number 88; all of its isotopes are radioactive.
Radon	Element of atomic number 86; it is a colorless, radioactive noble gas and the alpha decay daughter of radium.
RBE	Relative biological effectiveness.
RCRA	Resource Conservation and Recovery Act.
REAC/TS	Radiation Emergency Assistance Center/Training Site.
Rem	Roentgen equivalent man; a measure of equivalent radiation dose; dose in rads multiplied by a quality factor; 100 rem = 1 Sv (the SI unit).

RERF	Radiation Effects Research Foundation (Japan).
Risk, absolute	Excess risk due to exposure to a radiation.
Risk, relative	Ratio of the risk of those exposed to a radiation to those not exposed.
Risk coefficient	The increase in incidence or mortality rate per unit radiation dose.
Roentgen	A special unit of radiation exposure, absolute for X- and gamma rays up to 3 MeV; 1 R = 2.58 × 10^{-4} coulombs of electrically charged ions per kilogram of dry air.
RTG	Radioisotope thermoelectric generator.
Sarcoma	A malignant tumor arising from connective tissue.
SBWR	Small boiling water reactor.
SI units	International system of units; radiation units in this system include becquerel, gray, sievert.
Sievert	SI radiation unit; the product of grays and any modifying factor, 1 Sv = 100 rem.
SIR	Standardized incidence ratio.
SMR	Standardized mortality ratio.
Somatic	Pertaining to an effect limited to an individual as opposed to a genetic effect. The former affects only one person, the latter affects subsequent generations.
Specific activity	The activity of a radionuclide divided by the mass of that nuclide.
SPEERA	Secretarial Panel for the Evaluation of Epidemiologic Research Activities.
SRL	Savannah River Laboratory.
SRP	Savannah River Plant.
Stem cell	A generalized mother cell that on division differentiates into more specialized cells, such as B- and T-cell lymphocytes.
Stochastic	Denoting effects where the probability, but not the severity, is a function of dose without threshold; heredity effects or cancer incidence are examples of stochastic effects.

Stopping power	A measure of the energy loss of a charged particle as it passes through a material.
Superfund	*See* CERCLA.
T65	A system of gamma-ray and neutron exposures assigned to the bomb victims at Hiroshima and Nagasaki based on their distance from the bomb center and any shielding involved.
Teratological effects	Effects that cause grossly abnormal biological forms.
Thyroid	A two-lobed endocrine gland found in the neck of human beings; it produces the hormones thyroxin and triiodothyromine.
TMI	Three Mile Island, the site of a nuclear power reactor.
Trabeculation	Anchoring strands of connective tissue.
Transmutation	Any process in which the nucleus of an atom is changed into a different nucleus, usually of a different atomic number.
Transuranium	Refers to elements of atomic numbers greater than that of uranium, atomic number 92.
Tuff	A rock composed of compacted volcanic ash.
TVA	Tennessee Valley Authority.
^{235}U	A radioactive isotope of uranium of mass 235; it is fissionable.
^{238}U	A radioactive isotope of uranium of mass 238.
Unattached fraction	The fraction of airborne daughters of radon that are not yet attached to particles.
Uranium	Element of atomic number 92.
USCEA	United States Council for Energy Awareness.
USTUR	United States Transuranium Registry.
USUR	United States Uranium Registry.
Weighting factor	The ratio of the total stochastic risk resulting from radiation to a tissue to that of the total risk to the body when it is irradiated uniformly.
WHO	World Health Organization.
WIPP	Waste Isolation Pilot Plant, Carlsbad, NM.

Working level	Any combination of short-lived radon daughters in 1 L of air that can potentially release 1.3×10^5 MeV of alpha energy.
Working level month	The cumulative exposure consisting of the number of working levels received in a working month.
X-ray	Electromagnetic radiation of high energy, usually above 1 keV; X-rays are nonnuclear in origin.

References

The definition of a particular term can be found in many reference documents. For the societal and radiational sense of this book, the following two sources were found to be extremely useful.

1. Borders, R. J. *The Dictionary of Health Physics and Nuclear Science Terms;* RSA Publications: Hebron, CT, 1991.
2. *Dorland's Illustrated Medical Dictionary,* 27th ed.; W. B. Saunders Co.: Philadelphia, PA, 1988.

Author Index

Affiliation Index

Subject Index